Disputatio

An International Transdisciplinary Journal
of the Late Middle Ages

Volume 2: Constructions of Time
in the Late Middle Ages

1997

Disputatio

An International Transdisciplinary Journal
of the Late Middle Ages

Volume 2:
Constructions of Time in the Late Middle Ages

Edited by Carol Poster and Richard Utz

Northwestern University Press
Evanston, Illinois 1997

Northwestern University Press
Evanston, Illinois 60208-4210

Printed in the United States of America

ISBN 0-8101-1541-7

The paper used in this publication meets the minimum requirements of the American National Standard for Information Sciences -- Permanence of Paper for Printed Library Materials, ANSI Z39.48-1984.

Editorial Note

Disputatio is an international transdisciplinary annual of the late middle ages (ca. 1300-1550). Each issue provides exhaustive and diverse treatments of one significant example of late medieval culture by scholars representing various nations, approaches, and disciplines. We feature scholarly articles, inclusive review essays, and/or a comprehensive bibliography on the topic of each issue.

This volume shows the many ways in which time was constructed in late medieval society in response to a variety of intellectual and practical problems. On the one hand, philosophers considered time in terms of Christian theology and recently rediscovered Aristotelian physics, while on the other hand, sailors and farmers needed to measure hours and seasons for their daily tasks. People from all classes of society -- kings, peasants, merchants, poets, or clergy -- were profoundly affected by the new chronological technologies of the late middle ages. This volume reflects both the cosmopolitan nature of the late medieval world and the interdisciplinary nature of late medieval studies. It includes studies of Arab, Scandinavian, Spanish, English, Italian, and Latin Europe and the disciplines of literature, cultural history, philosophy, history of science, theology, and textual criticism.

We would like to thank the members of our advisory and editorial boards. Special thanks go to Martin Camargo, Bonnie Wheeler, Werner Wunderlich, the University of Tübingen Library, the English Department of University of Northern Iowa, and our reviewers.

Carol Poster and Richard Utz,
Editors

Contents

Chaucer's *Chronographiae*, the Confounded Reader, and Fourteenth-Century Measurements of Time

Peter W. Travis

Just a few hours out of London on the pilgrimage to Canterbury, Harry Bailly, the pilgrims' leader, suddenly decides to determine precisely what time of the day it is. He does this not by reading a portable mechanical clock, since these light-weight coiled-spring instruments were not invented until the fifteenth-century.[1] Nor does he determine that it was the hour of ten *ante meridian* by consulting what was called a "chlyndre," a portable dangling sun-dial such as that sported by the lecherous monk in "The Shipman's Tale."[2] Nor, as was most common in the Middle Ages, does he almost accidentally learn of the time by hearing bells rung from a tower in a nearby town or abbey. Rather, Harry uses his own extraordinary powers of discrimination to determine that, for this precise latitude, since the sun had run a bit more than a half hour more than the fourth part of its course through the artificial day, and since the shadow of every tree at this very moment was absolutely equal in length to the tree's height, and since the date of this day could be inferred or known to be the eighteenth day of April, and since the sun had ascended precisely forty-five degrees into the sky, it could readily be asserted that, beyond all shadow of a doubt, it was exactly "ten of the clokke." And by "clokke," a word first recorded in English in 1371,[3] Harry does not mean an old-fashioned sundial or waterclock but rather the newly-invented mechanical clock that indicates sixty-minute, or equinoctial, hours whose length and *termini* remained constant throughout the day and year.

This is an astonishing moment, not only in the entirety of Chaucer's literary corpus, but in all of medieval literature. To my knowledge, nowhere in Western literature before Harry's calculation does a literary figure determine the exact, equinoctial hour of the day. Rather, in all medieval literature until at least the last quarter of the fourteenth century, time remained essentially what it had always been: natural and cursive, hours changing in length as the seasons expanded and contracted through the agricultural and liturgical year.[4]

Harry's time-telling on the first day of the pilgrimage is in fact not the only moment when Chaucer's conventional markings of time are ruptured by a technically convoluted temporal periphrasis. A second instance occurs at the very end of the pilgrimage, just before the "Parson's Tale," when Chaucer the narrator-pilgrim determines that the sun is descended from the meridian not quite twenty-nine degrees, that a man's shadow is at this moment eleven times one-sixth the height of his body, that the zodiacal sign Libra is beginning to ascend above the horizon, and that it is precisely four of the clock. Such an egregiously complicated and technical time-measurement occurs in only one other place on the pilgrimage, this time inside a tale, namely, "The Nun's Priest's Tale." Chauntecleer, the tale's rooster-protagonist, determines the precise equinoctial hour of the day since he knows "by nature" the degrees of the sun's ascension; and the Nun's Priest, the tale's horologically challenged narrator, attempts to determine the month and date of that day via a calculation so convoluted that most readers refuse to go through the mental contortions necessary to arrive at anything close to the actual day upon which the events of this beast fable putatively happen.

There are several extraordinary things about these three passages. The first oddity is that they are very rarely discussed by scholars other than editors. Until very recently, the only two scholars to have paid sustained attention to the scientific technicalities of these passages, Sigmund Eisner and John D. North, are both astronomers by training.[5] But literary critics are not immediately to be faulted for their reluctance to engage in such scientific calculations, for the second unusual element concerning these time-telling episodes is that Chaucer's ideal contemporary interpreter had to know how to use either an astrolabe or a complex scientific almanac in order to corroborate the hour of the day, the date of the day, and the ascension of the sun at the Canterbury pilgrims' precise geographic latitude. Although a highly allusive and intertextual poet, Chaucer nowhere else in his poetic corpus requires that his readers hold a scientific text in hand in order to determine the accuracy and rightness of a passage of poetry. But the most challenging feature of all three *chronographiae* is also the most frustrating and confusing: when the scientifically inclined reader goes through all the necessary calculations to verify Chaucer's data, it turns out that each and every one of Chaucer's three determinations of time is somehow wrong!

In this essay I intend to examine Chaucer's three "wrong" *chronographiae* by calling upon both sides of the reader's bicameral brain. For humanists this means developing a high tolerance for scientific and philosophical lore: mathematical tables, physio-kinetic diagrams, metalinguistic word problems, an astronomical syzygy, and logical speculations about change, motion, and time, as well as about the kinematics of beginnings and endings. Because I am persuaded that the significance of Chaucer's *chronographiae* cannot be understood by merely reading them in a formalist fashion, any more than the significance of time can be comprehended by staring at the heavens (or a calendar, an astrolabe, a sundial, or a mechanical clock), I begin by considering five modes of understanding and representing time in the fourteenth century. While these five modes do not provide all the information necessary to assessing Chaucer's chronographic art, all five are called into play as Chaucer's readers attempt to determine that phenomenon we offhandedly call the time of day.

I: Five Medieval Ways of Looking at Time

One mode of time measurement in the Middle Ages was **computistical**. The *computus* was the science by which medieval scholars sought to forecast the precise date of Christ's resurrection and thus the precise dates of all the other movable feasts in the liturgical year. Although eight full days of Oxford University's liberal arts astronomy syllabus were given over to studying a dizzying array of tables, charts, diagrams, and mathematical formulas, and although all medieval Western Christians were in agreement that Easter day should be the Sunday after (but not on) the fourteenth day of the Paschal moon -- that is, the calendar moon whose fourteenth day, reckoned from the new moon inclusive, falls on, or is the next following, the vernal equinox -- in Chaucer's time the *computus* had proved strikingly inadequate. Nature, failing to conform to humanly prescribed divisions of sidereal, solar, and lunar movement, was obdurantly lifting into the fourteenth-century heavens a sun and a moon whose motions mocked all the authoritative tables of solar positions and lunar phases, as well as ushering in a true vernal equinox roughly a week before the orthodox date of March 21.[6] Irreducible even in the eighteenth century to a comprehensible algorithm,[7] the principles of mathematical and astronomical calculation for determining Easter day were in the fourteenth century a constant reminder of the extraordinary complexities and mysteries of mastering and measuring time.[8]

A second mode of analyzing time in the later Middle Ages was **philosophical**. The urgency of the most prominent fourteenth-century metaphysical speculations on time devolved from two major condemnations which had been pronounced by Bishop Stephen Tempier in 1277. In one condemnation Tempier anathematized the extreme Aristotelian proposition: "If heaven stood still, fire would not burn flax, because time would not exist." But in a second condemnation he evenhandedly anathematized the extreme Augustinian assertion: "Time and eternity have no existence in reality but only in the mind."[9] Because of Tempier's condemnations, fourteenth-century philosophers had to hypothesize an understanding of time which, although deriving from the works of their two great temporal authorities, Aristotle and Augustine, fully embraced the position of neither. Duns Scotus and Peter Aureole, for example, invented a phenomenon called "potential time" -- which inheres inside the mind when understood as a continuous succession without parts, but which inheres in some substance outside the mind when understood as "measured time," a composite of continuous or discontinuous arithmetical magnitude.[10] Ockham's extensive treatise on time was even more subtle. For Ockham, the word "time" like the word "motion" principally signifies the ninth heavenly sphere, the *primum mobile*, moving with maximum speed and uniformity; but the word "time" unlike the word "motion" also consignifies the soul numbering or observing that heavenly sphere. Ockham intermittently implies that the diurnal movement of the *primum mobile* is in fact the ultimate clock, but he just as regularly retracts this Aristotelian position by arguing that the very idea of regular movement is a human conception generated to conform to a purely ideal clock that our mind constructs as soon as it perceives any movement or change whatsoever. So where then does this ideal of temporal measurement, this Absolute Clock, exist? Ockham never satisfactorily answers this question. And neither do Walter Burley, John Buridan, Albert of Saxony, or Marsilius of Inghen, all of whom nevertheless attempt to determine the nature of the Absolute Clock, its relationship to the *primum mobile* as well as to the motions of our minds.[11]

The third salient mode of analyzing time was **mechanical**. Until the invention of a block-and-release mechanism, or escapement, which made use of oscillatory motion to divide time into countable beats, the nature of civilized time-keeping had been in the West essentially the sound of bells rung by human beings at certain unfixed times during the day in response to the movement of a shadow across a sundial. This was all dramatically modified by the mechanical clock, a late thirteenth-century invention which many historians believe constituted "one of the most important turning-points in the history of science and technology."[12] Despite the advent of mechanical clocks, the old ways of "knowing time" remained dominant, that is, either sensing time by an intuitive feel, or attending to monastery or church bells which measured the day into seven unevenly divided liturgical *horae*: lauds, prime, tierce, sext, none, vespers, and compline. But the appearance of so many magnificent and elaborate block-and-release clocks in the center of city squares -- such as the tower clock of Norwich Cathedral (1325), the St. Albans clock

(1330), and the famous de' Dondi clock of Padua completed in 1364 -- not only inspired further urban horological competition, but, in the judgment of Jacques Le Goff, effected a violent collision between the old and the new meanings of time.[13] Used in some cities as *Werkglocken* to regulate the length of working days no matter how long or short the natural day, these clocks, uniquely designed to measure equal and unchanging hours twenty-four hours a "day," created a new consciousness which Le Goff calls "merchant's time"-- where time is seen as something secular, measurable, mechanized, commodifiable, wasteable, exploiting as well as exploitable.[14] To balance Le Goff's vision of crisis, however, I think we must also emphasize that these clocks in their very designs expressed a social pride in humankind's ability to imitate and venerate the designs of divine creation: "The hourly parade of saints and patriarchs, the ponderous strokes of the hammer-wielding jacks, the angel turning with the sun, the rooster crowing at sunrise; the lunar disk waxing and waning with the moon -- all these movements and sounds," writes David Landes, "offered lessons in theology and astronomy to the up-gazing multitude that gathered to watch and wonder."[15]

The fourth way of determining time in the Middle Ages was **astrolabic**. As the author of the *Treatise on the Astrolabe*, a treatise which in North's view is "the first competent work in English on such a subject,"[16] Chaucer was clearly fascinated with the workings of one of the most remarkable instruments of the Middle Ages. A handheld brass instrument consisting of a graduated circle and a sighting rule which pivoted about a pin topped with a horse's head, the astrolabe allowed observations of the altitude of any star or planet. What enabled the astrolabe to serve as astronomical computer was its set of circular brass plates: the uppermost plate, the "rete," represented the rotating heavens and contained a star map and an eccentric circle representing the ecliptic; this was superimposed upon a second plate called the "climate," which bore the projection of a fixed coordinate system defined for the latitude of the user. This system consisted of an horizon line, circles of equal altitude, and lines of equal azimuth, as well as the celestial equator, tropic of cancer, and tropic of capricorn. The rete would be rotated over the climate in order to make a variety of calculations concerning such matters as the rotation of the heavens, the position of the sun, and the time of day. As Marijane Osborn has shown, Chaucer's fascination with this instrument may be seen in a number of places in the poetry of the latter part of his career, most notably in the magical steed of brass in "The Squire's Tale," which is fashioned as a larger-than-life model of an astrolabe.[17]

The fifth and final way of assessing time in the late Middle Ages was **kalendric**. It is this mode which is most important in the study of Chaucer's *chronographiae*, not only because Chaucer expects his most attentive readers to consult a kalendar, but because the kalendar incorporates into its measurements at least three of the four other modes of temporal analysis I have so far discussed. If on an English country road one did not have an astrolabe in hand and for some unnatural reason wished to know the precise time of day, what could one do? The answer is to be found in Chaucer's *Treatise on the Astrolabe*, where he cites two English contemporaries, John Somers and Nicholas of Lynn, each the author of a *kalendarium* -- that is, an ingenious and highly rationalized system of determining mechanical clock time without the use of mechanical clocks (which were notoriously inaccurate anyway). John Somer's *Kalendarium* (1380) and Nicholas of Lynn's *Kalendarium* (1386) both reveal the many different ways time could be reckoned in the fourteenth century. The latter is of more immediate relevance because it is this almanac which Chaucer relies upon in all three *chronographiae*.[18] It is the almanac, in other words, which Chaucer's readers had to know how to use *if* they wished to emulate, and verify, Chaucer's own calcul-ations of time. The following description of this *kalendarium* is thus meant to stand in for Chaucer's readers' direct experience of reading it -- of poring over its instructions, learning how to decipher its tables, corroborating its data with their own astrolabic calculations, and putting these corroborations to experiential use.

Nicholas of Lynn's *Kalendarium* opens with a month-by-month calendar from the year 1387 to the year 1462. Each page consists of a series of vertical columns. The left-most column, labeled *Numerus dierum*, contains a vertical line of numbers running from 1 to 28, 29, 30, or 31, depending upon the month. Next is a column of numbers ranging from 1 to 19, indicating each day's Metonic cycle (that is, the order in which the moon returns with near perfection to the position it held in the nineteenth year before the given year). Next follows a column for the Roman calendar, designating each date by ides and kalends. Next is a column assigning all the saints' days, after which follows a double column labeled *Verus locus solis*, indicating the degree and minutes of a degree of the sun in a given sign of the zodiac at noon for each day of the year. After this there follows a double column labeled *Quantitates diei artificialis*, which catalogues the length of each "artificial day," given in hours and minutes, from the point of the sun's rising above the horizon to the point of the sun's setting below the horizon. This is followed by a second double column labeled *Quantitates diei vulgaris*, the "vulgar" or "natural" day being the artificial day plus the morning and evening twilights.

For our purposes, the next important column is something unique to Nicholas of Lynn's *Kalendarium*, a shadow scale, in degrees and minutes, entitled "The altitudes of the sun and the lengths of the shadows of any man six feet high in hours of equal distance [i.e., hours as measured by a mechanical clock] at 51 degrees, 50 minutes," which is the latitude of Oxford, England. The column is in fact double, divided into *altitudines solis* and *umbre hominis*, so that, for instance, at 5 p.m. on October 1 the sun is 4 degrees high and the length of the shadow of a six-foot man is 85 feet and 51/60ths of a foot.

Following these pages of columns there is another section that includes a table of solar eclipses, lunar eclipses, the ascendant and beginnings of the celestial houses, the reign of each planet, a table of moveable feasts, tables tracing the motion of the sun and the dignities of the planets, and a chart indicating which zodiacal sign the moon is in each day of the year. In light of the many ways of measuring time represented in the *Kalendarium* -- solar, lunar, Roman, liturgical, astrolabic, and mathematical/scientific -- Nicholas's concluding manual placed at the very end of his almanac is all-important: it serves as a guide to the use of its disparate tables and points in the direction of deeper questions concerning the nature and measurement of time itself.

What is collectively indicated by this introductory survey of five medieval ways of assessing time -- computistical, philosophical, mechanical, astrolabic, and kalendric -- is the increasing awareness of the problematics of time developing within certain communities in the late fourteenth-century. Of course the majority of rural Europeans continued to live within the same time continuum as had their ancestors -- working and resting in harmony with the expanding and contracting daylight hours during seasons which were given sacred definition by the fixed and movable feast-days of the liturgical year. It was primarily in the cities that a self-conscious awareness of time's rationalized divisibility was taking hold: this fascination with measuring and meting out units of "mercantile" time is something that Chaucer's London audience was very much in the process of assimilating. But that part of Chaucer's potential readership which participated most fully in all five ways of assessing time was without question the scholarly community. In *Schools and Scholars in Fourteenth-Century England*, William J.Courtenay notes that "At the opening of the fourteenth century, students shared with their nonacademic contemporaries the same sense of time: solar time."[19] Like rural laborers, scholars measured their day into twelve liturgical hours which expanded to fill up the time from sunrise to sunset no matter the day's length. In northern latitudes this guaranteed that the length of classes and class-days fluctuated greatly with the seasons, the result being that a "lecture in early June in England might be almost twice as long as one given in mid-December."[20] However, by the end of the fourteenth century, university life in places like Oxford had been so affected by the mechanical clock that daily academic exercises were now measured in equal units of time. Thus in less than one century England's intellectual community had shifted the scheduling of its daily life from medieval solar time to modern clock time. In fact, Courtenay goes so far as to speculate that the great number of fourteenth-century Oxford treatises on the phenomenology of time may be traced directly to the influence of "one of the wonders of that age," the large mechanical astronomical clock at the monastery of St. Albans. Whichever way the influence worked, Courtenay continues,

> ... the development of the mechanical clock in England is contemporary with a growing fascination at Oxford with units of time and with measurement, which by 1340 had produced a mathematical *Physics* as well as a quantitative theology. The problem of the divisibility of temporal and spatial *continua*, which was a fundamental issue within the philosophy, science, and theology of fourteenth-century Oxford, may have caught the interest of so many because it corresponded to changes in their daily experience.[21]

As an intellectual attuned to the philosophical concerns of Oxford, as an authority on the technical and scientific intricacies of the astrolabe, and as a cosmopolitan civil servant working long but "regular" hours, Chaucer was clearly engaged in the many ways contemporary time could be measured and assessed. Without doubt, Chaucer's personal fascination with the different modes of ascertaining the time of day influenced his decision to include in the *Canterbury Tales* three state-of-the-art horological calculations. However, it is not so much the literary egregiousness of these technical calculations themselves but rather the unique way Chaucer presents them to his readers which needs most carefully to be explored. The majority of Chaucer's contemporary readers would of course never have troubled themselves with extra homework assignments to determine the right time of his literary days. And modern scholars of Chaucer, because of the shortcut "solutions" and correct "answers" supplied by editors of the *Canterbury Tales*, fail to experience anything close to the complicated *process* of intellectual inquiry that each of Chaucer's *chronographiae* invites and makes possible. It is only by simulating the analytic experiences of any one of Chaucer's well-educated, "clergial," and contemporary readers that the philosophical significance of these passages begins slowly to emerge. Working back and forth between Chaucer's text, Nicholas of Lynne's *Kalendarium*, and other modes of measuring and thinking about time, Chaucer's inquisitive reader runs into a series of puzzlements and intellectual cul-de-sacs. It is only after these scholarly frustrations modulate in the direction of a necessary but peculiar heuristic strategy that the confounded reader begins to make progress; that is, in due time, having contemplated the many errors in calculation presented in these *chronographiae*, and then having reflected at length upon the concept of "error" itself, Chaucer's reader will be able to arrive at a more sophisticated understanding of the temporal anxieties of the late fourteenth century.

II: Tempus est Hominis Confusio

Fragment II of *The Canterbury Tales* opens the "Introduction to the Man of Law's Tale" with Chaucer's first technical *chronographia*:

> Our Hooste saugh wel that the brighte sonne
> The ark of his artificial day hath ronne
> The ferthe part, and half an houre and moore,
> And though he were nat depe ystert in loore,
> He wiste it was the eightetethe day
> Of Aprill, that is messager to May;
> And saugh wel that the shadwe of every tree
> Was as in lengthe the same quantitee
> That was the body erect that caused it.
> And therfore by the shadwe he took his wit
> That Phebus, which that shoon so clere and brighte,
> Degrees was fyve and fourty clombe on highte,
> And for that day, as in that latitude,
> It was ten of the clokke, he gan conclude,
> And sodeynly he plighte his horse aboute....(1-15)[22]

"Our Hooste"'s means of arriving at the time of day is quite a feat for someone with neither astrolabe nor *Kalendarium* in hand, for, by some natural gift, he is able to make four inter-related calculations. First, he perceives that the artificial day, that is, the passage of time from the exact point of the sun's rising above the horizon to the exact point of the sun's setting below the horizon, is more than one-fourth gone. In fact, relying upon some interior chronometer, he has been able to divide the artificial day into quarters and then add the amount of time that has passed since the completion of that first fourth: the sun has "ronne/ The ferthe part, and half an houre and moore ...". The next calculation Harry makes is the equally extraordinary determination that the day's date is April 18 -- which, being a numerical designation, is also unusual in the

Middle Ages especially when made by one "nat depe ysterte in lore." Harry's third calculation is perhaps not so astonishing, for it is a mode of affixing "natural" time that can be traced at least as far back as Anaxagoras.[23] Although Harry apparently does not get off his horse and actually measure any shadows, he determines that because all shadows at this moment are equal in length to the objects that cast them, the sun "[d]egrees was fyve and fourty clombe on highte." The final measurement, time "for that day, as in that latitude,/ ... was ten of the clokke." And this clock we should be reminded, is a mechanical clock, measuring "equal," or 60-minute hours, rather than a sundial that measures "unequal hours."

Harry's temporal calculations are obviously informed by an expertise which no human being could master without the assistance of an astrolabe and/or *kalendarium*. Although most readers do not stop to ponder the impossibilities, it is obvious that one can not intuit, one can only learn of the measured length of the artificial day one is presently living in; one can not know, one can only discover from some text and/or instrument how many hours, at a specific moment, have passed beyond the first fourth of that artificial day; without the use of an astrolabe, one can not with absolute certainty ascertain that the sun is precisely forty-five degrees ascended; and, finally, if one were to use the less precise, shadow-scale method of measurement, without the help of a *kalendarium* one could not translate its natural hours into regular clock-hours.

Now, if we do choose to check Harrry's measurements by turning to the tables in Nicholas of Lynn's *Kalendarium* for the month of April, we see that Harry has got some matters quite right. April 18, for instance, is the only time in the entire month when at a regular hour (actually, at both 10 a.m. and 2 p.m.) the sun is precisely forty-five degrees in the sky, and thus the shadow of a six-foot man (or tree) is six feet long. But even so, as Sigmund Eisner has noted, Harry's first, "astrolabic" calculation is not entirely accurate. For according to Nicholas of Lynn, the artificial day of April 18 (that is, from sunrise [4:47 A.M.] to sunset [7:13 P.M.]) is 14 hours and 26 minutes long. One fourth of this artificial day, 3 hours and 26 minutes, yields 8:23 a.m.; add another "half an houre" and we arrive at 8:53 -- which is still a rather long time away from 10 a.m. So what does Harry, or Chaucer, mean by "and moore" in "half an houre and moore"? Eisner, feeling that "and moore" has to "signify an inappropriately long stretch to bring the time to 10:00 A M.,"[24] concludes that this faulty computation is not Chaucer's, but the Host's, who was "nat depe ystert in loore." However, Eisner is pleased to report that Harry's shadow calculation is right, not only because it agrees with the *Kalendarium*, but because he, Eisner, "standing in the rare sunshine of Chaucer's island and latitude," took the same shadow measurement with the same results, although it had to be taken on April 27, which is the Gregorian equivalent to Chaucer's Julian April 18.[25]

Eisner's explanation of Harry's scientific errors is based on his analysis of Harry's character: Harry is not very well educated. Nevertheless, throughout the pilgrimage Harry suffers from a very modern obsession with the passage of time, an obsession that is an instantiation of the mercantile mentality characterized by Le Goff. Harry is thus very much a man of the hour, epitomizing the burgeoning bourgeois fascination with the process of gaining and losing the measurable *minutiae* of time. Yet what is to be made of Harry's intuitive ability to measure and enumerate these *minutiae*? It is as if a calculator, astrolabe, clock, and *kalendarium* had all been programmed into the microchip of Harry's brain. Harry thus remains an unusually strained poetic conceit, his *a priori* abilities to measure clock-time "by nature" placing him in a peculiar relationship to the realistic surface of Chaucer's fiction.[26] And just as peculiar is his relationship to Chaucer's readers, who are invited to make their own calculations. While most readers will countenance Harry's time-telling powers as so much poetic license, a few "clergial" readers, conversant as Chaucer was with contemporary almanacs, will take it upon themselves to look into a *kalendarium* -- checking out the relevant columns, making the appropriate calculations, and double-checking their arithmetic -- only to discover, finally, that Harry's calculations were wrong. Although the fiction of Harry's improbable powers provides an invitation to duplicate Harry's calculations, the reader's reward for dutifully undertaking such extratextual research is to arrive at a small but significant miscalculation. But whose miscalculation is it? Harry's? Chaucer's? Nicholas of Lynn's? The reader's? Or two, or more, of the above?

To address the significance of this miscalculation will ultimatly involve addressing the issue of error itself. At the moment, however, it is most fruitful to move away from this one scene in Chaucer's pilgrimage and to consider the second instance of technical measurement of "non-fictive" time taken on the road to Canterbury. In the "Prologue to The Parson's Tale," near the very end of the pilgrimage, the narrator informs his readers in a roundabout way that it is getting late in the afternoon:

> By that the Maunciple hadde his tale al ended,
> The sonne fro the south lyne was descended
> So lowe that he nas nat, to my sighte,
> Degrees nyne and twenty as in highte.
> Foure of the clokke it was tho, as I gesse,
> For ellevene foot, or litel moore or lesse,
> My shadwe was at thilke tyme, as there

> Of swiche feet as my lengthe parted were
> In sixe feet equal of proporcioun.
> Therwith the moones exaltacioun--
> I meene Libra--alwey gan ascende
> As we were entryng at a thropes ende .... (1-12)[27]

Although very much a part of the same tradition, the perspective of Geoffrey's *chronographia* is quite different from Harry's. Whereas Harry's provides a secular measurement of quotidian time *in medias res*, this vision of time is numinous with signs of finality, judgment, and death. The twenty-nine degrees of the descending sun may symbolically renumerate the twenty-nine original tale-telling pilgrims; the impossibly long and carefully measured shadow cast by the narrator's idealized six-foot body seems profoundly spiritual in its demarcations;[28] the sun's descending, the moon's rising, and Libra, the scales of judgment, suspended above the pilgrims' entrance into a nameless town's end, all suggest time's yearning for eternity. The eschatological tenor of these images has in fact led one scholar, Russell Peck, into the far reaches of numerological allegory: for Peck the 4 of "four of the clokke" is the "number of Fortune, a number which is intimately associated with the world"; 29 is the number of "transgression, decrepitude, sterility, and concupiscense"; 11 is also a number of sin and transgression; while 6 is a number of the perfect soul and a sign of earthly perfection.[29]

Were the text not so obviously concerned, even here at the end of the pilgrimage, with the scientific measurement of human time in the here-and-now, it would be tempting to elevate all of these arithmetical calculations far away from the literal and into typological, tropological, and mystical *allegoresis*. But the empiricalness of these matters and their obvious obsession with the accuracies of scientific measurement simply cannot be denied -- even though the narrator's measurements are even more innacurate than were the Host's. The first difficulty is that while *The Riverside Chaucer* prints "foure of the clokke," all manuscripts but one (MS Ch) give the time of day as "ten of the clokke." If "ten of the clokke" has to be a scribal error (as all Chaucer's editors presume), its correction appears rationalizable in one of two ways. Eisner argues that since the Roman sign for ten was an x and the Arabic sign for four was an x with an arc over the top, modern editors are justified in emending a mistaken ten to four.[30] Peck, from a different angle, argues that Chaucer is using one of the several modes of hourly numeration employed in the Middle Ages: "ten of the clokke" means ten hours after liturgical prime; and since prime (Peck presumes) is equivalent to 6 a.m., the tenth hour of the day is, actually, four p.m. But even after this textual crux (if it is such) is amended, two other difficulties remain. At 4 P.M. in mid-April, as Nicholas of Lynn's *Kalendarium* makes absolutely clear, it is Saturn, rather than Libra, which is the moon's exaltation. How could Chaucer, or Geoffrey, his narrating alter-ego, manage to get this very basic piece of information so wrong? The other egregious time-telling problem concerns the day itself. Unlike Harry, Chaucer does not provide the date of the day whose hours he is measuring. So it is left to the reader to turn once again to Nicholas's *Kalendarium*, and then to focus upon the *altitudines solis* column and the *umbre hominis* column for the month of April, and eventually to determine that the only possible date for the last day of the Canterbury pilgrimage is April 17. That is, on April 17 the sun's altitude at 4 P.M. is 28° 57' -- or, "nat, to my sighte,/ Degrees nyne and twenty as in highte." And the length of a six foot man's shadow at 4 P.M. is ten feet ten inches (51/60) long -- or, "ellevene foot, or litel moore or lesse." Thus, working with both Chaucer's and Nicholas of Lynn's texts, readers and editors must eventually conclude that April 17 is the last day of Chaucer's pilgrimage to Canterbury.

But determining the day's date, rather than settling all matters, only succeeds in creating more time-telling problems: as we recall, the Host's *chronographia* has already determined that the first day of the pilgrimage was April 18. By these calculations, the Canterbury pilgrimage -- setting out from London on the morning of April 18 and nearing Canterbury on the afternoon of April 17 -- ends one day before it begins. What is the reader to make of this paradox? In the prototypical medieval literary journey, the *Divine Comedy*, it makes perfect sense that time at a certain point should leap backward: that Dante, moving from the northern to the southern hemisphere as he leaves Inferno and arrives at the base of Purgatorio, should actually gain twelve hours, the result being that he is able to live through the same Holy Saturday for a second time in the same year. But in Chaucer's earthbound pilgrimage within the same latitude, unless time is itself moving backward (a phenomenon Stephen Hawking says is scientifically feasible at certain phases of world history[31]), the chronically earthbound interpreter is left, once again, in a quandary. Eisner's solution to all these difficulties is to proclaim that they are "not significant" because Chaucer uses these dates only for "symbolic purposes." The one symbolic purpose Eisner advances is this: "[T]he tone of the Parson's Tale is fitting for Good Friday. The joyous arrival in Canterbury would be fitt ing for Easter. In 1394 Good Friday was 17 April."[32] Such "symbolic" explanations are attractive, and as one form of literary exegesis need not be "wrong," but in my judgment they nevertheless refuse to confront the issue of "error" head-on, for surely Chaucer could have used a symbolic mode which did not get tangled in such convoluted scientific conflicts.

What, therefore, are we to do with this plethora of arithmetical and astronomical miscalculations? First, even though the purpose of the resulting confusion is not entirely clear, I think we must acknowledge that Chaucer was aware he was inscribing errors of measurement into his *chronographiae*: these are errors of authorial intention, not of authorial ignorance or carelessness. Second, given that Nicholas of Lynn's diverse tables and explanatory canons provide encyclopedic instruction in the many ways time has been and may be measured, we must recognize that Chaucer has invited his careful readers to become equally careful readers of the *Kalendarium*. However, what Chaucer expects his readers finally to glean from their analysis of Nicholas of Lynn's various tables and from Chaucer's own "misreading" of these tables requires further thought. Third, despite the similarities between the two *chronographiae* examined to this point, it is clear that they embody two quite different relationships to time. Time for Harry is anxiously concentrated in the *hic et nunc* of the absolute present, whereas time for the Chaucerian narrator is both a phenomenon to be measured and a mystery to be pondered *sub specie eternitate*. The dialogue between these two different perspectives needs to be further articulated. Finally, unless we are able to accept Harry's uncanny time-telling instincts as merely so much poetic license, we need to consider how his superhuman powers raise questions not only about the narrator's presumed omniscience in telling "fictive" time but also about anyone's ability to know scientifically what the "right" time really is.

Together, these chronographic issues begin to suggest the possibility that a coherent inquiry may underly Chaucer's exploration of the human interpretations of fourteenth-century time. In themselves, however, they do not provide adequate answers to the questions they obliquely raise. To more closely approximate these answers, we now must consider the most complex *chronographia of The Canterbury Tales*, that found in the middle of "The Nun's Priest's Tale." To appreciate the intellectual context of the Nun's Priest's time-telling puzzle, it will be necessary to trace out a very specific set of medieval logical studies pertaining to temporal beginnings and endings. But because of certain interpretative difficulties deeply embedded in the Nun's Priest's *chronographia*, a knowledge of these studies of time, motion, and change will (once again) not provide any "right" answers. Rather, the Nun's Priest's *chronographia* will have succeeded in foregrounding a set of conceptual questions which prove, finally, to be the backbone of his pedagogical agenda: *understanding* the problematics of temporal calculation and *understanding* the significance of not arriving at right answers are what give ultimate coherence to Chaucer's explorations of the mysteries of time.

III: The Nun's Priest's *Sophisma De Incipit Et Desinit*

More than half way through "The Nun's Priest's Tale," the narrator suddenly takes it upon himself to let his readers know when his fable putatively happened:

> Whan that the monthe in which the world bigan
> That highte March, whan God first maked man,
> Was complet, and passed were also
> Syn March bigan thritty dayes and two,
> Bifel that Chauntecler in al his pryde,
> His sevene wyves walkyng hym bisyde,
> Caste up his eyen to the brighte sonne
> That in the signe of Taurus hadde yronne
> Twenty degrees and oon and somwhat moore,
> And knew by kynde and by noon oother loore
> That it was pryme, and krew with blisful stevene.
> "The sonne," he seyde, "is clomben up on hevene
> Fourty degrees and oon, and moore ywis.... " (4377-4389)[33]

In this passage, the rooster Chauntecleer determines "by nature" that it is the liturgical hour of prime. And he is able, relying again on his native knowledge (like Harry he needs no astrolabe), to determine that the sun is forty degrees "and more ywis" in the sky. The narrator supplies the information that the sun has ascended twenty-one degrees "and somwhat moore" in the sign of Taurus. With only this amount of information, the dutiful reader could go straight again to Nicholas of Lynn's *Kalendarium* and read down the first column until Taurus appears -- to discover that the sun is in Taurus 21° 6' ("twenty degrees and one, and somwhat moore") on May 3. Further, at 9 a.m., which Eisner says is equivalent to "pryme," the sun is at an altitude of 41° 17' ("Fourty degrees and oon, and moore ywis"). And, if it matters at all, one could add that the shadow of a six-foot man (or six-foot chicken) at this moment is six feet ten inches long. Finally, because 152 lines later the Nun's Priest notes in an aside that "on a Friday fil al this meschaunce," one could perhaps assert, as Eisner has done (using the *Kalendarium* once again), that the precise Friday on which "The Nun's Priest's Tale" occurs was either May 3, 1381 or May 3, 1392. For once, then, all the parts seem to fit, and Chaucer, the Nun's Priest, Chauntecleer, and the reader finally get everything right.[34] However, in the opening four lines of the *chronographia* the narrator attempts to determine the month and day in quite a different way, and these lines complexify the reader's determination of the proper time of the tale's events:

> *Whan that the monthe in which the world bigan*
> *That highte March, whan God first maked man,*
> *Was complet, and passed were also*
> *Syn March bigan thritty dayes and two,*
> Bifel that Chauntecler in al his pryde ...

Considering the unusual degree of syntactic and semantic complexity in this linguistic collocation, and its ambiguous overlappings of beginning points and ending points in time, it is not surprising that these four italicized lines have succeeded in generating the longest single gloss in the *Variorum* edition of "The Nun's Priest's Tale" -- a 152-line history of five centuries' worth of scribal emendations, editorial amendations, critical interpretations, and scholarly exegesis. It is also perhaps not surprising that almost every modern edition of Chaucer has gone entirely against the authority of all extant manuscripts and rewritten the phrase "Syn March bigan" as "Syn March was "gon." Although we shall see eventually that these four lines parody a specific kind of medieval word-problem dealing with instants of beginnings and endings in time, the responses of scholars unaware of the tradition are nevertheless instructive in dramatizing the heuristic scope of the passage. The first point that needs to be underscored is that all readers are forced to follow the same general path at the outset. After first re-reading the opening four lines several times, then muttering quietly to themselves a version of the fifteenth-century mnemonic poem --

> Thirty dayes hath November,
> April, June, and September;
> Of XXVIII is but oon,
> And all the remenaunt is XXX and I [35]

-- then counting carefully on their fingers (for, having got this far, it would be embarrassing to be wrong), and then wondering at least subliminally whether it makes any difference to measure from the ending of one unit of time (such as a day) or from the beginning of another unit of time (such as the next day), scholars have nevertheless managed to arrive at more than one apparently correct determination of the date Chaucer presumably "had in mind." Three of the most popular dates proffered have been April 1, April 2, and May 2, but the all-time favorite among Chaucerians is Chaucer's favorite day, the third day of the third month, May 3.[36] Yet even those who are certain their May 3 reading is right admit that the grammatical and syntactical construction of this sentence is extremely tortuous and even perhaps self-contradictory.

For many of Chaucer's scribes, editors, and readers, the way of solving these linguistic problems has simply been to correct or bracket the most irritating, illogical, or erroneous-looking parts of the sentence. The scribes of certain, less well-known, manuscripts were the first to take it upon themselves to straighten things out, by changing "thritty dayes and two" to their notion of the proper number of days that should have passed by. Thus "thritty dayes and two" becomes "thritty dayes or two" (MS S1); "sway monthes and two" (MS Ha); "sevyn dayes and two" (MS He); or "thritty dayes and mo" (MSS Ph and Ry).[37] Because these scribal emendations of the number of elapsed days create as many arithmetical problems as they seek to solve, various editors have taken it upon themselves to change, or reinterpret, what is clearly the crucial verb: "bigan" in the subordinate temporal clause "Syn March bigan." Thomas Tyrwhitt (1775) boldly emended "bigan" to "ended." Francis Thynne (1599), Andrew Brae (1870), and Mark Liddel (1930) all chose to interpret "bigan" as meaning "be gonne" as in "gone by or departed." Thus "Syn March began" actually means "March having passed by."[38] Robert Pratt in his 1974 edition confidently emends "bigan" to "was gon"[39]; likewise, in *The Riverside Chaucer* (1987), Larry Benson replaces "bigan" with a bracketed "was gon," hypothesizing that originally "Chaucer wrote *was gon* and that scribal recopying from [line] 3187 led to the mistaken repetition of *bigan* in [line] 3190."[40] The result of these decisions is that the majority of students now reading Chaucer will find:

> Whan that the month in which the world bigan,
> That highte March, whan God first maked man,
> Was compleet, and passed were also,
> Syn March [was gon], thritty dayes and two,
> Bifel that Chauntecleer in al his pryde ...[41]

One rare exception to this spate of editorial emendations is the *Variorum* edition of "The Nun's Priest's Tale," where Derek Pearsall allows "Syn March bigan" to stand unchanged. Although he concedes the meaning of the passage "is not transparently clear at first sight," Pearsall is nevertheless persuaded that "the text is clearly capable of unambiguous interpretation, and does not need emendation."[42] Pearsall's authority proves to be none other than William Skeat, who had contended in 1894 that "[t]he words 'since March bigan' are parenthetical" (which means, apparently, deletable or non-significant); thus, Skeat concludes, "we are, in fact, told that the whole of March, the whole of April, and two days of May are done with."[43]

At this point I wish to make a major assertion: Chaucer intended these lines to be written as they have been written. Any editorial correction is in fact a very grave error, because Chaucer's educated contemporaries would all have instantly recognized these four lines as a clever parody of one of the most persistent and popular intellectual problems in medieval logic and *Physics*. Logically, this problem dealt with two verbs, "to begin" (*incipire*) and "to cease" (*desinire*). In formal propositions, these two verbs often proved to be ambiguous regarding their scope; they were seen to contain covert references to the past or the future despite their being in the present tense; and they were seen to involve covert negations despite their affirmative forms. As several historians of medieval philosophy have shown,[44] *incipit* and *desinit* were included in every late medieval logical textbook analyzing syncategorematic words -- that is, those crucial little words in a proposition which determine the syntactical, rather than semantic, operations of a sentence, even though, properly speaking, as verbs they are *categorematica*, i.e., either nouns or verbs.[45] For example, consider this rather straightforward twelfth-century illustration of what is known as the logical "fallacy of division":

> Those two men cease to be.
> If anyone ceases to be, he dies.
> Therefore those two men die.[46]

The fallacy of the inference is this: since the two men as a pair may cease to be by way of the death of only one of them, it is erroneous to infer that both of them die. A more complicated form of logical scrutiny was represented in treatises which focused upon the fact that under certain conditions *incipit* and *desinit* involve covert negation and covert references to the past and for future. Consider the fallacious inference in the next sophism:

> Socrates begins to be white.
> Therefore he begins to be colored.[47]

The covert negation and the covert temporal assertion are clearly interconnected: that is, in positing "Socrates begins to be white" the verb "begins" is also positing that immediately before now he was *not* white. The fallacy of the inference is that having been not white immediately before the present is not incompatible with having been some other color.

What moved *incipit* and *desinit* into the center of scholarly attention in the thirteenth and fourteenth centuries was the discovery of Aristotle's *Physics* and the gradual recognition that Aristotle's complex philosophy of limits -- concerning the nature of the limits of motion, change, and time -- was peculiarly receptive to a logico-linguistic re-application within the receptive framework of fallacious syllogisms focusing upon the syncategorematic verbs of beginning and ceasing. In Book 6 of the *Physics*, Aristotle maintains tht since time is a continuous phenomenon, temporal instants, rather like geometrical points on a geometrical line, are not actually real but are merely theoretical entities. Although time is infinitely divisible into instants, time is not made up of instants -- which means that between two instants there exists an infinitude of further divisible instants. According to Aristotle, any change from one state to another must occur instantaneously, within a hypothetical instant of time; therefore, he writes, "that which has changed must at the moment when it has first changed be in that state to which it has changed" (*Physics* 6.5 235b6). In other words, the basic question is whether or not there *is* a first moment of change, or what Aristotle calls a "primary when." Common sense would assert that when a ball rolls down a ramp, there is a first instant of motion, call it T1, which follows immediately upon a last instant of non-motion, call it TO. But Aristotle, hewing close to his fundamental belief in the infinite divisibility of time, would answer that between T1 and TO there are an uncountable number of instants, and that for any given instant of change there exists a prior instant of change. Thus, he writes, "it is evident that every thing which is in motion must have moved before" (*Physics* 6. 6. 236b32). Equally important for Aristotle is the question of whether there is a first element of *completion*, an instant when it is true to say "this change is over." Aristotle's answer to this limit decision problem is that, although one must deny that there is no "primary when" at the beginning of continuous change, one must nevertheless affirm there is a "primary when" signifying that that change has been completed.

The interpretation by medieval thinkers of Aristotle's position on limits was complicated by his treatment of another kind of change considered elsewhere in the *Physics*. In Book 8, Aristotle contemplates what he calls "contradictory change," that instant when some subject changes within a given time interval from one state to another, such as from being non-white to being white. The question is, what is the status of the subject at the instant of change? Aristotle's answer, in contrast to that reached in Book 6, is that the instant of change belongs to the time when the subject is white, rather than non-white. What this means, then, is that with contradictory change, and in direct contrast to continuous change, it is true to say there is a first moment when something *is* white, yet there is no last moment when one can say it is *not* white.

In attempting to develop a single doctrine that would unify these two passages in Aristotle, medieval thinkers decided that, in denying a "primary when" at the beginning of continuous change, Aristotle was denying a first instant of the existence of that change. Next, they determined that, in asserting a "primary when" at the completion of continuous change, Aristotle was in fact asserting a first instant of the non-existence of that change. What this interpretation led to was a fairly complex definition of the nature of the beginnings and endings of continuous change. Since one must deny the existence of a first instant of a continuous change, one must perforce maintain that there is a last instant *before* the beginning of that change; likewise, since one must affirm the first instant of the change no longer existing, one must perforce deny the existence of a last instant of the change itself. What this means is that every continuous change is *extrinsically* limited at both ends: it is limited at its beginning by its last instant of not-being and it is limited at its end by its first instant of not-being. A diagram supplied by the most ardent and informative student of these issues, John Murdoch, illustrates the extrinsic limits of continuous things, or *res successiva*: (see Figure 1):

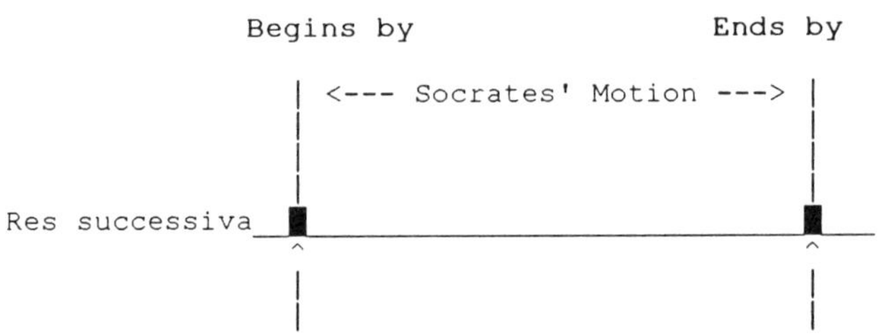

(1) Last instant of non-being
(2) Extrinsic limit
(3) *Remotio* of the present, *positio* of the future

(1) First instant of non-being
(2) Extrinsic limit
(3) *Remotio* of the present, *positio* of the past

Figure 1: *Res successiva*

But, concerning contradictory change, matters were perceived by medieval logicians to be quite different, in light of Aristotle's definition of such change in Book 8 of the *Physics*. Contradictory change (such as Socrates' being white) differed from continuous change (such as Socrates running) in that white was a *res permanens*, a permanent thing, all its parts existing at once, whereas Socrates' running was held to be a *res successiva*, its parts existing one after the other. It will be recalled that Aristotle maintained that the moment of change from non-white to white was contained within the time of his being white, which for the scholastics meant that there was a first instant of that change. Unlike continuous change, then, contradictory change is limited *intrinsically* at its beginning. But even though there is a *first* instant of that contradictory change, we recall that Aristotle allowed as to how no *final* instant of that state could be indivisibly posited. Therefore scholastic thinkers asserted that contradictory change at its ending was limited *extrinsically* (just like continuous change) and that its external limit was the first instant of its non-being.

As scholastic interpretations of Aristotle's limit-decision positions developed, a number of treatises were written which focused upon the physical implications, or object-language analysis, of these matters, including Walter Burley's "De instant" (c. 1340), John of Holland's "Tractatus de primo et ultimo instant" (1369), and Peter of Mantua's (d. 1400) *De instanti*.[48] More popular, however, were those studies which remained essentially linguistic in nature, situating Aristotle's temporal theories into the pre-existing logical tradition of *incipit/desinit* syncategorematic analysis. These studies included William of Sherwood's (d. 1272) *Treatise on Syncategorematic Words*, Peter of Spain's (d. 1277) even more influential *Tractatus syncategorematum*, and Thomas Bradwardine's (d. 1349) *Tractatus de incipit* et *desinit*.[49] By far the most important permutation of the Aristotelianized *incipit/desinit* phenomenon appeared in the second quarter of the fourteenth century and flourished for the next century in such works as Richard Kilvington's (d. 1361) *Sophismata*, William Heytesbury's *Regule solvendi sophismata* (ca. 1335) and Paul of Venice's (d. 1492) *Logica magna*.[50] This phenomenon was the rage for inventing irresolvable logical sophisms, or *sophismata*, wherein all the complicated issues of limit decision cases, including the discrimination of permanent versus successive things, were encapsulated in a highly compressed sophistical construction. In his survey of this form of limit-decision studies, Norman Kretzmann has counted almost 250 different examples of late medieval *incipit/desinit sophismata*.[51] Here, by way of illustration, is one quite complex sophism posed by Chaucer's brilliant contemporary William Heytesbury:

> Assume that Socrates is now one foot tall, Plato two feet tall; and let each of them grow uniformly throughout the next hour, Socrates twice as fast as Plato, stipulating further that three feet is the smallest size that neither of them will have, since both cease to exist at the end of the hour at which moment both would have been three feet tall had they then existed.[52]

Although students of philosophy would have spent hours attempting to resolve a word problem such as this, the major point in their study was not to simply find a solution -- because more often than not no solution was possible. In fact, time and again students discovered that a final and correct answer had been thwarted or immobilized by a logical impasse -- something called in formal logic *vis confundendi*, and which I translate informally as the "stymie principle."[53] Immobilizing the reader in a quandary, the unresolvable sophism is designed to force the reader into further and deeper rumination both upon linguistic and logical issues at hand, as well as upon phenomenological matters dealing with beginnings, endings, motion, change, and time.

With all of this at least generally in mind, let us now return to Chaucer's crucial four lines:

> Whan that the monthe in which the world bigan
> That highte March, whan God first maked man,
> Was complet, and passed were also
> Syn march bigan thritty dayes and two ...

How might well-educated medieval readers have read this? In accordance with their scholarly training in logic, they would first break down the sophism into its *exponentes*, or exponible parts.[54] The *exponentes* are two premises and one implied inference:

> The month of March has passed.
> Thirty-two days have passed.
> Therefore: A total of sixty-three days have passed.

Written out this way, the sophism's ambiguities are obvious, since it is perfectly possible that thirty-one of the thirty-two days which have passed are contained inside the month of March itself. What gives special edge to the sophism is Chaucer's combining its two premises with the innocent-looking conjunction *and*: "and passed were also ..." One of the major words addressed in all medieval syncategorematic textbooks was "and" (*et*) -- because it is an ambiguous delimiter, capable of serving either as a divisive conjunction or as a composite conjunction. For example, consider *and*'s radically different meanings in these two sentences: "My son and my friend are here" versus "My son and my friend is here." And consider the rules that make it allowable to say "two and three are five" but never "two are five and three are five." The pressure Chaucer places upon the ambiguity of *and* is dramatized equally well by *also*, which is *and*'s intensifier in the phrase "*and* passed were *also*." As in Modern English, so in Middle English, *also* could be used as a simple additive, meaning "in addition, sequentially." However, it could be used with equal clarity of meaning to signalize a substitution for a preceding statement, meaning "likewise, concurrently." Thus we have the resulting ambiguity which appears to afford two equally right readings. One is sequential, adding thirty-two days to the end of March. The other is concurrent, adding thirty-two days to the beginning of March. The alternate syllogism is therefore:

> The month of March has passed.
> Thirty-two days have passed.
> Therefore: A total of thirty-two days have passed.

Because of the ambiguous nature of the two key syncategorematic words "and" and "also," two different readings seem to be equally available: sixty-three days have passed, or thirty-two days have passed. But are both readings equally valid? Clearly they are not. The second term of the syllogism is not simply "Thirty-two days have passed," but rather "Thirty-two days have passed *since March began* [passed were also/ Syn March bigan thritty dayes and two]." Therefore the two passages of time -- concurrent rather than sequential -- are in fact the *same* passage of time (plus one day), measured first in a month and then in days. Therefore, the valid solution to the *incipit/desinit* part of the Nun's Priest's *chronographia* cannot be either May 2 or May 3.[55] Rather, if there *is* a valid solution to the sophism (and of course there may not be), it has to be either April 1 or April 2.

But which is it? The sophism's evocation of medieval *incipit/desinit* scholarship would seem to promise that an application of that tradition would produce a definitively right answer. Having dealt with one *incipit* ("Syn March *bigan*") by using a quite straightforward logico-linguistic analysis, we are justified in hoping that the sophism's several beginnings ('Whan ... the world *bigan*," "when God *first* maked man," "Syn March *bigan*") and ceasings ("*Was complet*," "*passed* also"), as well as its various mixtures of different units of time (days, months, and perhaps years) could all be sorted out. Following good neo-Aristotelian thinking, what we find, however, is that time, at least in this case, cannot be neatly sorted out. A preliminary step in the medieval logical thinking process would be to determine that all these units of time are phenomena of continuous change rather than convertible change, *res successiva* rather than *res permanens*. Then, taking the precept which holds that every unit of continuous change is limited extrinsically at both ends, we see that the limits of the month of March are actually outside (rather than inside) the month of March. In other words, in order to measure from that point when "March bigan," we must measure from the last point before March actually becomes March; and in order to measure from the end of March, when it "was complet," we must measure from the first point when March is no longer March. Next, we must consider the termini of those two crucial days, Day 31 and Day 32 ("thritty dayes and *two*"). Day 31, as we have seen, has to be March 31, and thus it clearly occurs inside the month of March. However, the *incipit* of Day 32 (April 1) must also occur *inside* the month of March, because the beginning-point of every unit of time is extrinsically limited. What this leads us to is the seemingly absurd proposition that the *incipit* of April actually precedes the *desinit* of March 31. In other words, March ends in April, but April begins in March!

But how could it be that a *subsequent* unit of time (April) actually *begins* before the *preceding* unit of time (March) *ceases*? How could the beginning of the new day precede the ending of the old day? Our careful reasoning about time -- according to the best medieval authorities -- leads to a daunting and perhaps irresolvable paradox.[56] Let us put it one more way: if we measure 32 days from the *beginning* of March, we end up at the last instant inside April 1. However, if we measure one day from the *ending* of March, we end up at the first instant inside April 2. Thus the time of "The Nun's Priest's Tale," its fictive present, seems to hover somewhere between the ending of one day (April 1) and the beginning of the next (April 2). But what is most weird is that this unit of time which we have determined to call the fictive present exists as an instant *before the ending of the past* and *after the beginning of the future*. While a resolution to this strange paradox remains a possibility, I leave its achievement to others, for it should by now be clear that Chaucer's major point is heuristic: that is, his parodic *incipit/desinit* sophism is meant first and foremost to raise questions about the nature of time and the complexities of its mensuration, rather than to posit a date upon which "The Nun's Priest's Tale" happens.

Thus, for Chaucer and his learned contemporaries, determining the *philosophical incipit*-point or *desinit*-point of sequential temporal units was obviously no easy matter. However, determining the *conventional* termini of any "non- philosophical" unit of time, such as an "everyday-day," was *also* fraught with difficulties. As Linne Mooney has observed, there were five methods used in the Middle Ages to posit a day's *terminus a quo* (all five are incorporated into the *kalendaria* of John Somers and Nicholas of Lynn): the day could begin at midnight; it could begin at sunrise; it could begin at noon of the preceding day, it could begin at sundown of the preceding day; or, very importantly, it could begin at noon of the *stated* day.[57] Since the astronomical day was generally understood to begin at noon of the *stated* day (this, at any rate, is Nicholas of Lynn's practice), the forenoon of each calendar day actually belonged to the preceding astronomical day. It therefore follows in "The Nun's Priest's Tale" that the forenoon of April 2 (the calendar day) is actually contained within the diurnal confines of April 1 (the astronomical day).[58] Is it possible, then, that Chauntecleer escapes the jaws of death on the morning of the day before the day he escapes the jaws of dcath? Unless Chaucer's time is once again moving backwards, it may ultimately be best to recognize another "stymie principle" at work and accede to yet another paradox, namely, that the historical date of "The Nun's Priest's Tale" may in fact be two days, *both* April 1 *and* April 2. Indeed, this paradox may simply be a variation of the "philosophical" paradox arrived at above which places the tale between the ending of April 1 and the beginning of April 2. But even if this ecumenical solution were entirely satisfactory, there is still the matter of "March." When Chaucer specifies "March" as "the month in which the world began," is he in fact saying that Chauntecleer's fall from grace actually occurred immediately after the very first month in which the world was created? Even though "literary" readers are unlikely to be troubled by this difficulty in semantic supposition, it is precisely this kind of *vis confundendi* that Chaucer's logical readers would expect to find embedded in an *incipit/desinit* sophism. Whereas our preceding analysis of "began," "was complet," "and," and "also" focused on the sophism's key *syncategorematic* words, the ambiguities of "March" shift our attention to the sophism's most crucial *categorematic* word. It takes little training in logic to appreciate that "March," standing alone, could denote any month by that name in history, and it likewise takes little training to see that the doubly restrictive identification of this March as "the month in which the world bigan, ... when God first maked man" pulls its referential scope powerfully in the direction of the very beginning of human time. Of course, what happens syntactically is that all three possible meanings of March play off against each other in the reader's mind: March is the proper name of the first month at the beginning of time; March is the common name of the first month of any and every year; and March is the proper name of the unique month of that specific year in which "The Nun's Priest's Tale" happened. Although some scholars, undaunted by Chaucer's sophistical springs, have confidently identified that year as 1392,[59] using the precepts of medieval logic should lead us away from such confidence. Rather, it is irresolution, paradox, ambiguity, and confusion that are the most important consequences of the heuristic strategies of Chaucer's brilliant *incipit/desinit* sophism. It is very doubtful, therefore, that Chaucer ever expected his careful reader to determine the sophism's "right" day-- April 1 or April 2[60] -- and it is just as doubtful that

Chaucer ever expected that date to be successfully harmonized with the later date the remainder of the *chronographia* points toward -- May 2 or May 3. Finally, even though it is extremely doubtful that Chaucer ever expected his reader to locate his beast fable in any specific year of human history, he certainly makes plausible the implausible inference that these events happen very near the beginning of time.

But, then, at what point does time begin? No matter where we turn, it appears that the recurring question "What time is it?" continues to convert itself into Augustine's most famous question, "What *then* is time?"

IV: Incommensurability in Chaucer's Time

Because he believed that time past and time future exist only in our subjective perceptions in the present, Augustine despaired that we could ever know the true nature of time. The only way we could possibly come to understand time is if we could somehow grasp the *In principio*, that mysterious moment where Eternity touches the *incipit*-point of time's beginning. Augustine's dream of reaching time's point of origin seems to have been realized by Dante the pilgrim in *Paradiso* 29. Arriving at the *primum mobile*, the ninth sphere of the universe, Dante observes the sun and the moon poised in perfect oppositional alignment, bisecting the earth by the plane of the same horizon. As Alison Cornish has argued, in pondering this extraordinary syzygy Dante is also observing the "primary when" of creation, that instant at the very beginning of time when the angels were created and when they fell.[61] Chaucer, in contrast, shares none of Augustine's anguish or Dante's hubris concerning the human desire to achieve a privileged *point d'appui* from which to view time's mysteries. Instead, Chaucer reminds us, while grounded in the *hic et nunc* of the present, we are nevertheless empowered with quite powerful disciplines with which to measure the *minutiae* of time. These disciplines, although never fully in harmony with each other, embrace the five modes of assessing time outlined earlier -- computistical, philosophical, mechanical, astrolabic, and kalendric. Of course, the human desires which our application of these disciplines may seek to satisfy are quite varied. As we have seen, the Host's calculations in the first *chronographia* on the pilgrimage are driven by his anxieties concerning the irrecoverable expenditure of time, and Geoffrey's calculations at the end of the pilgrimage are informed by a human need to penetrate through time into the realm of eternity. In the last *chronographia* we examined, that of "The Nun's Priest's Tale," it is the desire for a philosophical understanding of time which is most pronounced. This philosophical desire is foregrounded by Chaucer's careful use of formal logic, the liberal art which the later Middle Ages considered to be its most precise discipline for analyzing any phenomenon, be it theological, metaphysical, or physical. While the Nun's Priest's sophism invokes an extremely refined set of logical instruments for analyzing time, it also demonstrates how the thoughtful application of these instruments may nevertheless lead into uncertainty and confusion.

Chaucer's mixture of scientific professionalism, logical sophistication, and intellectual skepticism concerning our ability to understand time is unique in late medieval literature: there is simply nothing comparable to his technical *chronographiae* and the way they obliquely instantiate the epistemological fissures caused by the impact of the mechanical clock. Nonetheless, Chaucer's general perspective on the human parameters of "knowing" time is strikingly similar to the position advanced by one of the fourteenth century's greatest scientists, Nicole Oresme.[62] Oresme's magnum opus, *Tractatus de Commensurabilitate Vel Incommensurabilitate Motuum Celi* (ca. 1351-1362), examines an issue integral to the nature of time itself: the commensurability or incommensurability of the motions of celestial bodies. If the ratio between the motions of the celestial bodies is a rational number, Oresme explains, then the cosmos is controlled by a clear principle of order: as a consequence of this order, exact calendars and almanacs could be composed; future astronomical conjunctions could be confidently predicted; and the notion of the Great Year, in which all stars return to their original position, would be ratified. However, if the ratio between some of these celestial bodies is an irrational number, they are then incommensurable: in a cosmic system of even partial incommensurability, exact calendars and almanacs could not be composed; future astronomical conjunctions could not be predicted; and the Great Year would prove to have been a great myth. The first two books of the *Tractatus* constitute an extensive, highly technical, and fairly balanced defense of the potential validity of each of the opposing positions; the third book, which takes the unexpected form of an allegorical dream-vision, is designed as a resolution to the debate. In this book, a dreamer-mathematician brings his concerns to Apollo, who, as "ruler of the Muses and Sciences," summons Arithmetic and Geometry to teach the dreamer what he wishes to know. But these two figures immediately start quarreling. Arithmetic, who espouses commensurability, argues that "[i]t seems unworthy and unreasonable that the divine mind should connect the celestial motions ... in such a haphazard relationship."[63] Geometry, who champions incommensurability, finds a more beautiful pattern in a combination of regular and irregular ratios: "an harmonious union of them is more excellent than a separate [and independent] uniformity."[64] At the end of this unresolved disputation, Apollo feels "adequately informed"; the dreamer, however, finds himself "astonished and confounded." Although Apollo promises to provide the definitive answer once and for all, the dreamer reports what happens next: "alas, the dream vanishes, the conclusion is left in doubt, I am ignorant of what Apollo, the judge, has de-creed on this matter."[65] On this note of disappointment, the *Tractactus* comes to an abrupt end.

Oresme's unsatisfactory ending is of course deliberate, for the entire *Tractatus* is premised upon Oresme's principled conviction that every form of scientific inquiry is limited, imperfect, and subject to unavoidable error. Even though his sympathies clearly lie on the side of incommensurability, Oresme repeatedly underscores the mere *probability* of each and every thesis: in fact, he allows, "nothing prevents certain false propositions from being more probable than certain true ones."[66] Thus, rather than providing a definitive answer to its fundamental question, Oresme's *Tractatus*, as Steven Kruger notes, "focuses attention again and again on the problem of attaining knowledge."[67] The frustrations built into the *Tractatus* eventually lead Oresme's confounded dreamer to cry out: "But, oh immortal gods who know all things, why did you make the very nature of men such that they desire to know, and then deceive or frustrate this desire by concealing from us the most important truths?"[68] But Oresme himself actually finds the existence of inscrutability cause for celebration and hope: "it would be better that something should always be known about [celestial revolutions], while, at the same time, something should always remain unknown, so that it may be investigated further."[69]

Chaucer's strategies in his three *chronographiae*, I contend, are similar to Oresme's. Each *chronographia* encourages rigorous scientific inquiry, but each eventually throws into doubt the possibility of any verifiably "truthful" conclusion. Each requires the reader to integrate several of the liberal arts (logic, astronomy, geometry, and arithmetic), while implying that the use of these liberal arts will not help us arrive at any form of certitude. Each seems to promise the ardent scholar some kind of intellectual revelation at the very end, only to make that ending deeply disappointing. While Chaucer's stymied reader is quite justified in registering various forms of frustration, it is clear that Chaucer's use of the *vis confundendi* principle is meant to be enabling and instructive. The "wrong answers" embedded in these *chronographiae* are designed neither to satirize intellectual inquiry nor to foster scientific paralysis. Precisely because right and satisfactory solutions cannot be found to the problems of mensuration posed in the *chronographiae* -- not only can they not be found, they in all likelihood do not exist -- when readers are confronted by these *insolubilia* they are forced to switch the focus of their attention to the very *process* of intellection itself. Certainly, a great deal of the reader's critical interrogations happens in the process of working through each calculation, as so many aspects of the ways we measure time are engaged, explored, and scrutinized. But just as much intellection may follow the completed investigation, when, having attained a certain level of knowledge, the reader is able to think more independently about our conceptualizations of time -- whether anxiously like the Host, spiritually like Geoffrey, or philosophically like the Nun's Priest. In the end, the most remarkable feature of these *chronographiae* is that they are essentially auto-didactic exercises: in themselves they "say" little, because Chaucer grants the reader full freedom to fill in the interpretive spaces they make available. Ultimately, then, what we need to acknowledge about Chaucer's *chronographiae* is not that they fail to achieve closure but that they so brilliantly succeed in their open-endedness. This open-endedness is promising: one of these days we may indeed arrive at an adequate understanding of time, a matter about which none of us, as yet, is "depe ystert in lore."

Dartmouth College

Notes

1. See David S. Landes, *Revolution in Time: Clocks and the Making of the Modern World* (Cambridge: Harvard UP, 1983), 86.

2. For a depiction of the cylinder dial, see J.D. North, *Chaucer's Universe* (Oxford: Oxford UP, 1988), 112.

3. Linne R. Mooney, "The Cock and the Clock: Telling Time in Chaucer's Day," *Studies in the Ages of Chaucer* 16 (1993), 103.

4. The use of the term "cursive" I owe to Anne Higgins's useful study, "Medieval Notions of the Structure of lime," *Journal of Medieval and Renaissance Studies* 19 (1989), 227-58; In "The Hours of the Day in Medieval French," *French Studies* 13 (1959), 240-51, W. Rothwell notes that "right up to the end of the fourteenth century indications of time in vernacular texts can only rarely be translated at all accurately into precise clock hours," and that "One of the most time-conscious writers of that age, Froissart, ... changed from the old cononical hours to the new clock hours in the middle of his *Chroniques*" (p. 249).

5. Professor Eisner informs me that his knowledge of astronomy has been acquired entirely through self-instruction. With the exception of Mooney's "The Cock and the Clock," an essay concerned with modes of measuring time rather than the difficulties of Chaucer's *chronographiae*, studies of Chaucer's temporal paraphrases have tended to be allegorical, rather than scientific, in their allegiances. See, for example, Chauncey Wood, *Chaucer and the Country of the Stars* (Princeton: Princeton UP, 1970), 70-102; 161-72; 272-97. See also the numero-

logical readings of Dolores Warwick Frese, *in An Ars Legendi for Chaucer's Canterbury Tales: A Reconstructive Reading* (Gainesville: U of Florida P, 1991), 141-63.

6. As late as 1532, Rabelais could still satirize the outrageous effects of the computists' calendar reckonings: "In that year the Kalends were fixed by the Greek date-books, the month of March was outside Lent, and mid-August fell in May. In the month of October, I believe, or perhaps in September -- if I am not mistaken, and I want to take particularly care not to be -- came the week so famous in our annals, that is called the Week of Three Thursdays." François Rabelais, *Pantagruel*, trans. J. M. Cohen (Harmondsworth: Penguin, 1955), 171. (Cited by Higgins, 234.)

7. See Gordon Moyer, "The Gregorian Calendar," *Scientific American* 246 (May 1982), 146.

8. For detailed studies of the computus in the Middle Ages, see A. Cordoliani, "Contribution à la littérature du comput au Moyen Age," *Studi Medievali*, ser. 3, vol. 1 (1960), 107-38; vol. 2 (1961), 169-208; J.D. North, "The Western Calendar -- 'Intolerabilis, horribilis, et derisibilis': Four Centuries of Discontent," in G.V. Coyne, M.A. Hoskin, and O. Pederson, eds., *Gregorian Reform of the Calendar* (Vatican: Pontificia Academia Scientiarum & Specolo Vaticana, 1983), 75-113; Faith Wallis, "Images of Order in the Medieval Computus," in Warren Ginsberg, ed., *Ideas of Order in the Middle Ages* (Binghampton, N. Y., 1990), 49-65. Arno Borst, *The Ordering of Time: From the Ancient Computus to the Modern Computer*, trans. Andrew Winnard (Chicago: U of Chicago P., 1993); and Charles W. Jones, ed., Bede, *Bedae Opera de temporibus* (Cambridge: The Mediaeval Academy of America, 1943), Introduction, 3-122.

9. For a sustained analysis of the effects of these two condemnations as well as a still enlightening discussion of fourteenth-century philosophical debates on the nature of time, see Pierre Duhem, *Medieval Cosmology: Theories of Infinity, Place, Time. Void, and the Plurality of Worlds*, Roger Ariew ed. and trans. (Chicago: U of Chicago P, 1985), 295-363.

10. Duhem, op. cit., 342-52.

11. Duhem, op. cit., 352-62.

12. Joseph Needham, as quoted (but not cited) by Landes, *Revolution in Time*, 53.

13. For a survey of the mechancial clocks known to have existed in late-fourteenth-century England, see Mooney, "The Cock and the Clock," 104-108.

14. See "Merchant's Time and Church's Time in the Middle Ages" and "Labor Time in the 'Crisis' of the Fourteenth Century From Medieval Time to Modern Time," in Jacques Le Goff, *Time. Work, and Culture in the Middle Ages*, trans. Arthur Goldhammer (Chicago: U of Chicago P, 1980), 29-52.

15. Landes, *Revolution in Time*, 79.

16. North, *Chaucer's Universe*, p. 38.

17. Marijane Osborn, "The Squire's 'Steed of Brass' as Astrolabe: Some Implications for the *Canterbury Tales*," eds. Patrick J. Gallacher and Helen Damico, *Hermeneutics and Medieval Culture* (Albany: SUNY Press, 1989), 121-31. Although it appears as a life-size magical toy which "in the space of o

day natureel" will "Beren youre body into every place," this steed of brass, Osborn shows, is clearly an astrolabe, the horse's head of which is operated by turning a pin: "when yow list to ryden anywhere,/ Ye mooten trille a pyn, stant in his ere," the Squire explains.

18. See *The Kalendarium of Nicholas of Lynn*, ed. Sigmund Eisner, transs. Gary Mac Eoin and Sigmund Eisner (Athens: U of Georgia P, 1980) both for the text of this *kalendarium* and for a full description of its parts and uses.

19. William J. Courtenay, *Schools and Scholars in Fourteenth-Century England* (Princeton: Princeton UP, 1987), 3.

20. Ibid., 7.

21. Ibid., 9.

22. Geoffrey Chaucer, *The Riverside Chaucer*, ed. Larry Benson (Boston: Houghton Mifflin, 1987), 87.

23. The ancient use of shadow scales is discussed briefly by North, *Chaucer's Universe*, 243. The use of the shadow-scale in medieval computistic/world maps is noted briefly by Evelyn Edson, 'World Maps and Easter Tables: Medieval Maps in Context," *Imago Mundi* 48 (1996), 41, appendix 6.

24. Eisner, *Kalendarium*, 31.

25. Ibid., 30 26. As J. D. North dryly remarks, "It is not easy to imagine even scholarly pilgrims in real life exchanging pleasantries about the fraction of the arc of the artificial day traversed by the Sun" (North, *Chaucer's Universe*, 123).

27. *The Riverside Chaucer*, p. 287.

28. Chaucer's shadow, to my knowledge, is the first human shadow to be recorded in English literature. Theoretically deduced from an almanac, it needs to be compared to the shadow cast by Dante's living body in Purgatorio: "Low at my back, the sun was a red blaze; its light fell on the ground before me broken in the form in which my body blacked its rays" (*Purgatorio* 3, 16-18). Mark Peterson observes that Dante's shadow is "oddly precise and geometric," giving evidence that "Dante must have spent a little time examining shadows carefully." See Peterson, "Dante's *Pysics*," in Guiseppe Di Scipio and Aldo Scaglione, eds., *The Divine Comedy and The Encyclopedia of Arts and Sciences* (Philadephia: John Benjamins, 1988), 167. According to Philip R. Berk, Dante's is the first human shadow found in all of medieval literature; see Berk, "Shadows on the Mount of Purgatory," *Dante Studies* 97 (1979), 47.

29. Russell A. Peck, "Number Symbolism in the Prologue to Chaucer's Parson's Tale," *English Studies* 48 (1967), 205-15.

30. Eisner, *Kalendarium*, 32-3.

31. Stephen W. Hawking, *A Brief Historv of Time: From The Big Bang to Black Holes* (New York: Bantam, 1988), 143-53.

32. Eisner, *Kalendarium*, 33.

33. Geoffrey Chaucer, *A Variorum Edition of the Works of Geoffrey Chaucer. Volume II. The Canterbury Tales, Part Nine, The Nun's Priest's Tale*, ed. Derek Pearsall (Norman, Oklahoma, 1984), 204-9. A recent study needs to be added to Pearsall's history: F.J.J. Peters, "Chaucer's Time in 'The Nun's Priest's Tale'," *Studia Neophilologica* 60 (1988), 167-70.

34. Admittedly, there is one small problem: the equation of the liturgical hora "pryme" with 9 a.m. is slightly willful, for as *The Riverside Chaucer* glossary reveals (1279), "prime" had at least three meanings for Chaucer: it could mean the canonical hour of 6 a.m.; or the first full hour of prime [6 to 7 a.m.]; or the three hours from six to nine in the morning. But in Chaucer's corpus, only "fully pryme," or "pryme large," means 9 a.m. In light of the heavy emphasis I have placed on the many other "errors" of temporal measurement in Chaucer's *chronographiae*, this matter, at least for the moment, will be allowed to pass.

35. *Middle English Lyrics*, eds., Maxwell S. Luria and Richard L. Hoffman (New York: Norton, 1974), 109.

36. See *The Riverside Chaucer*, 939, n. 3187--90.

37. John M. Manly and Edith Rickert, *The Texts of the Canterbury Tales* (Chicago: U of Chicago P, 1940), 7, 588-689.

38. All cited by Pearsall in *Variorum ... The Nun's Priest's Tale*, 205-6.

39. Robert Pratt, ed., *The Tales of Canterburv* (Boston: Houghton Mifflin, 1966), 244.

40. Benson, *The Riverside Chaucer*, 939, n. 3190.

41. *The Riverside Chaucer*, 258.

42. Pearsall, *Variorum ... The Nun's Priest 's Tale*, 206.

43. Quoted by Pearsall, *Variorum ... The Nun's Priest's Tale*, 206.

44. The two major studies of this tradition are John E. Murdoch, "The Analytic Character of Late Medieval Learning: Natural Philosophy without Nature," in *Approaches to Nature in the Middle Ages*, ed. Lawrence D. Roberts (Binghampton: Medieval and Renaissance Texts and Studies, 1982), 171-213, with "Comment" by Norman Kretzmann, 214-20; and Norman Kretzmann, "*Incipit/Desinit*," in *Motion and Time; Space and Matter: Interrelations in the History of Philosophy and Science*, eds. Peter K. Machamer and Robert G. Turnbull (Columbus: Ohio State UP, 1976), 101-36.

45. The fifteenth-century Middle English poem *The Court of Sapience*, ed. E. Ruth Harvey (Toronto: U of Toronto P, 1984), includes a tour of the Castle of Knowledge where each of the seven liberal arts is housed in a discrete apartment. The apartment of Dame Dialectica gently satirizes the most important logical topics and strategies of her time (63):

> Her parlour fresshe, her clothying proude and stout,
> Of *dyffert*, *scire* and of *incipit*
> With sophysmes full depeynted was aboute,
> And other maters, as of *desinit*.
> The *Comune Treatys* taught she them therwyth:
> Whiche, whatkyns, what is proposicioun
> What thyng he is, and his dyvysyon ...
> She taught hem there with lust and all lykyng;
> Fast they dysputen in theyr comonyng,
> With sophyms strong straunge maters they discusse,
> And fast they crye oft, *'Tu es asinus!'*

Although *The Court of Sapience* hardly reveals a profound understanding of dialectics, it expects its readers to recognize the contemporary logical buzzwords, among the most prominent being the syncategorematic terms *incipit* and *desinit*. This passage from *The Court of Sapience* is cited by John Murdoch, "The Development of a Critical Temper: New Approaches and Modes of Analysis in Fourteenth-Century Philosophy, Science, and Theology," *Medieval and Renaissance Studies* 7 (1978), 67.

46. Example provided by Kretzmann, "*Incipit/Desinit*," 105.

47. Example provided by Kretzmann, "*Incipit/Desinit*," 107.

48. For an excellent introduction to these tracts, see Curtis Wilson, *William Heytesbury: Medieval Logic and the Rise of Mathematical Physics* (Madison: U of Wisconsin P, 1966), 32-7.

49. Wilson, *William Heytesbury*, 38-40. Peter of Spain's discussion of *incipit* and *desinit* has been translated by Norman Kretzmann as an appendix to "*Incipit/Desinit*," 122-8.

50. Wilson, *William Heytesbury*, 29-56.

51. Kretzmann, "*Incipit/Desinit*," 133, n. 22.

52. Quoted by Wilson, *William Heytesbury*, 47; also cited by John Murdoch, "The Analytic Character of Late Medieval Learning," 193.

53. Some of the functions of *vis confundendi* performed in *incipit /desinit* sophismata are explained by Wilson, *William Heytesbury*, 29-56.

54. See Wilson, *William Heytesbury*, 11-2.

55. Considering the enormous investment of the critical establishment in the numerological/allegorical significance of May 3, it is not likely that the emendation of this sacred date will happen very soon. For a review of many of the ingenious meanings critics have discovered in Chaucer's May 3, see Pearsall, *Variorum ... The Nun's Priest's Tale*, 206-8.

56. While I find no solution in sight, I do not claim that none is available. A sophistical proposition which shares a number of similarities with Chaucer's sophism can be found in Heytesbury's *Regule solvendi sophismata*. Here is Curtis Wilson's presentation of the proposition, as well as its solution -- as supplied by Gaetano di Thiene (d. 1465): "Now there begins to be traversed some part which for the time ending at the present instant is as a whole still to be traversed. The proof proceeds as follows: (a) Now no part is traversed which for the time ending at the present instant is as a whole still to be traversed; this statement is obvious. (b) But immediately after the present instant there will have been traversed some part which for the time ending at the present instant is as a whole still to be traversed; for there will be no instant after the present instant such that prior to that instant some (indeterminate) part will not have been traversed. From (a) and (b) the required conclusion follows" (Wilson, *William Heytesbury*, 63).

57. Mooney, "Cocks and Clocks," 96.

58. But this is not the convention used in John Somers's *Kalendarium*. The result is Somers's forenoons are all a day before Nicolas's: all the evidence in Chauntecleer's calculations (if one were to use Somers's *Kalendarium*) verify May 2 as the right date, whereas in Nicholas's *Kalendarium* May 3, as we have seen, is the "correct" day.

59. For example, J.D. North is certain that 1392 is the only year possible (the tale's events must fall somewhere between the Peasants' Revolt and

Chaucer's death, he reasons); and the only concordant date within this year is Friday, 3 May, 1392; see North, *Chaucer's Universe*, 457-8.

60. Chaucer's literary and cultural environment could also have provided a debate between April 1 and April 2. Although April 1 in the fourteenth century had not yet acquired the burlesque features of April Fool's Day, it was nevertheless a holiday which enjoyed a touch of topsy-turvydom; it is perhaps appropriate, that the Nun's Priest warns in his *moralitas* that his tale may be taken as a "folie." April 2, on the other hand, was admired in the Middle Ages for its symmetry as the second day of the second month: in Machaut's "sit dou Lyon," for example, a poem which Chaucer may have translated as "the book of the Leoun," the crucial date is "Dou mois d'averil le jour secont" (see *The Riverside Chaucer*, 965, n. 1087). Thus, if Chaucer's readers wished to dispute this April 1/April 2 dialectic exclusively in regard to their cultural preferences (consider the equally "inconsequential" literary debate on the virtues of the flower and the leaf), there would be partisans on both sides.

61. Alison Cornish, "Planets and Angels in *Paradiso* XXIX: The First Moment," *Dante Studies* 108 (1990), 1-28.

62. That Oresme was "one of the fourteenth century's greatest scientists" has been asserted by Steven Kruger, *Dreaming in the Middle Ages* (Cambridge: Cambridge UP, 1992), 219, n. 87.

63. Nicole Oresme, *Nicole Oresme and the Kinematics of Circular Motion: Tractatus de commensurabilitate vel incommensurabilitate motuum celi*, ed. and trans. Edward Grant (Madison: U of Wisconsin P, 1971), 289-91.

64. Oresme, *Tractatus*, 313.

65. Ibid., 323.

66. Ibid., 311.

67. Kruger, *Dreaming in the Middle Ages*, 145.

68. Oresme,*Tractatus*, 287.

69. Ibid., 321.

Zeitkonstruktion als Kampf- und Protestmittel: Reflexionen über Joachim's von Fiore Trinitätstheologische Geschichtskonstruktion und deren Wirkungsgeschichte

Thomas Gil

ABSTRACT

How individual agents and social groups conceive time and space depends upon the material experiences they have and on how they interpret them using the categories and conceptual schemes which make sense in the socio-cultural in contexts in which they live. The following paper exemplifies what this exactly means, analysing the way Joachim of Fiore elaborated his temporal construction of a trinitarian world history and how Joachim's temporal construction was used as a critical conceptual instrument in three very different contexts: inside the Franciscan Order, by the German preacher Thomas Müntzer, and by philosopher F.W. Schelling.

Die im Mittelalter vorherrschende Geschichtsanschauung[1] läßt sich als eine temporale Konstruktion beschreiben, in der Geschichte als "heilsöko-nomische" Veranstaltung Gottes mit den Menschen aufgefaßt wird, welche aus zwei heilsgeschichtlichen Äonen oder Epochen besteht: dem alten Äon, der mit der Schöpfung anfängt, durch die Herrschaft des Gesetzes gekenn-zeichnet ist und durch den Sündenfall sowie dessen Wirkungsgeschichte maßgebend bestimmt wird, und dem neuen Äon, in dem die Gnade Gottes im Erlösungshandeln Jesu Christi manifest geworden ist. Gegen diese orthodoxe, hegemonische schöpfungstheologisch und christologisch-soteriologisch fundierte bzw. argumentierende Geschichtssicht opponieren zwei temporale Konstruktionen, welche Resultat bestimmter Erfahrungen und Erwartungen sowie ihrer kognitiven Verarbeitung sind: eine spiritualisierende, die Geschichte radikal auf Zukunft hin öffnende Zeitauffassung, die in der trinitätstheologischen heilgeschichtlichen Konstruktion des Abtes Joachim von Fiore ihren Ausgangspunkt und ihre maßgebende Bezugsgröße hat, und eine naturalisierende nüchterne Sicht vom Geschichtlichen, die eher implizit als explizit von den sogenannten "lateinischen Averroisten"[2] vorgetragen und verteidigt worden ist resp. in ihren Werken und Äußerungen enthalten ist. Beide Konstruktionen stellen Abweichungen von der herrschenden Sicht dar, die sich als Strategien der Kritik am Vorhandenen, Geltenden deuten lassen.

Die spiritualisierende Zeitauffassung betrachtete das Zeitalter seit Christus nicht als das letzte vor dem Weltende, sondern erwartete von der Zukunft eine epochale Wende zu einem neuen Zeitalter des heiligen Gei-stes, in dem eine vollkommenere Gestalt des irdischen religiösen Lebens Wirklichkeit werden würde. Diese spiritualisierende Zeitauffassung hat jahrhundertelang kirchenpolitisch und überhaupt politisch revolutionierend gewirkt, von den franziskanischen Spiritualen bis zu den Täufern des 16. Jahrhunderts. Die naturalisierende Sicht des Geschichtsverlaufs ergab sich aus der auf Aristoteles und Averroes zurückgehenden philosophischen Lehre, die vom ewigen Bestand der Welt ohne Anfang und Ende und von der dauernden Wiederkehr des Gleichen ausging.[3] Diese Lehre war mit dem christlichen Glauben an das Weltende ebenso unvereinbar wie mit dem Glauben an die Weltschöpfung. In den politischen Auffassungen und Staatstheorien der Schüler des Siger von Brabant, die ganz unbekümmert um Glaubensinhalte und kirchliche Festlegungen nach den natürlichen und rationalen Gesetzmäßigkeiten und Kräften des politischen Handelns fragten, wurde sie besonders wirksam.[4]

Im folgenden steht die "typologisch" vorgehende, spiritualisierende Zeitkonstruktion des Abtes Joachim von Fiore im Mittelpunkt. Sie wird als eine "kritische" Geschichtsanschauung gedeutet. Ihr kritisches Potential soll anhand einiger Einzelfälle ihrer Wirkungsgeschichte veranschaulicht werden. Die Geschichtskonstruktion Joachims von Fiore öffnet die Geschichte radikal auf Zukunft hin, auf eine bessere Zukunft, eine Zukunft "reicherer Gnade", indem sie drei Weltzustände unterscheidet und den gegenwärtigen Weltzustand des Sohnes in den kommenden Weltzustand des Heiligen Geistes münden läßt: einen Weltzustand der vollkommenen Erkenntnis, der Liebe, der Freiheit, der reicheren Gnade und Freundschaft. Diese trinitätstheologische Geschichtskonstruktion (Gegenstand des zweiten Abschnittes) hat eine komplexe, zum Teil revolutionäre Wirkungsgeschichte gehabt, die hier im einzelnen nicht rekonstruiert werden kann.[5] Vielmehr beschränke ich mich im dritten Abschnitt darauf, drei Stationen dieser Wirkungsgeschichte zu thematisieren und deren spezifische Besonderheit kurz zu charakterisieren. In einem ersten Abschnitt soll aber vorher im allgemeinen angedeutet werden, wie temporale Konstruktionen entstehen, worin deren Plausibilität besteht und welche Funktionen sie erfüllen können.

1. Geschichtliche Erfahrungskontexte und temporale Konstruktionen

In seiner Spätschrift *Die Krisis der europäischen Wissenschaften und die transzendentale Phänomenologie* führt der Philosoph Edmund Husserl den Begriff der "Lebenswelt" ein, um jenen ursprünglichen Kontext zu be-zeichnen, in dem Menschen im konkreten Lebensvollzug basale Erfahrun-gen machen und die "primordialen Bedeutungsschichten" von Wörtern entstehen, ohne welche Begriffe gar nicht gebildet werden können und die von Wissenschaftlern, wenn diese ihre Theorien entwickeln, selbstverständlich vorausgesetzt werden. In der "Lebenswelt" lernen die Menschen bestimmte Wörter sinnvoll zu gebrauchen; in ihr entstehen die fundamentalen Sinnbedeutungen, welche menschliche Kommunikation allererst möglich machen; in ihr haben auch, so die These Husserls, die wissenschaftlichen Grundbegriffe ihre ursprüngliche Plausibilität. Husserl redet von den "Ursprungsevidenzen" vorwissenschaftlicher Art der Lebenswelt, die für jede menschliche Verständigung im allgemeinen und für die wissenschaftliche Begriffsbildung im besonderen eine basale Bedeutung haben.[6] Husserl thematisiert allerdings nicht den geschichtlichen und sozialen Charakter der Lebenswelt. Dies ist auch der Grund, warum er den Begriff in der Singularform verwenden kann. Der ursprüngliche Erfahrungskontext der Lebenswelt ist aber durch und durch geschichtlicher und sozialer Natur. Deshalb gibt es nicht nur eine Lebenswelt, sondern verschiedene Lebenswelten, in denen sich unterschiedliche Sinnbedeutungsschichten von Wörtern, Begriffen und Diskursen bilden.

Entsprechend den Erfahrungen, die die Menschen in den verschiedenen "lebensweltlichen" sozialen und geschichtlichen Interaktionszusammenhängen machen, sind sie auch in der Lage, bestimmte Begriffe zu bilden und zu gebrauchen, die nur in solchen Interaktionszusammenhängen ihre volle Plausibilität haben, bzw. bestimmte Diskursarten zu entwickeln, die nur in solchen Handlungskontexten verständlich sind. Entsprechend ihren "lebensweltlichen" Erfahrungen, die immer auch Zeiterfahrungen sind, entwickeln die Menschen temporale Verhältnisse, d.h. spezifische Zeitbezüge, in denen die Vergangenheit, die Gegenwart und die Zukunft in einer bestimmten Art erlebt und bewertet werden können und faktisch auch bewertet werden. Diese praktischen, konkreten Zeitbezüge oder Zeitverhältnisse sind die Basis für die temporalen Konstruktionen, die in Form von Zeitauffassungen oder Geschichtstheorien dann vermittelt werden. Die temporalen Konstruktionen können mehr oder weniger komplex sein. Sie haben aber immer eine intersubjektive, kollektive Geltung und sind Orientierungsmedien für die an sie glaubenden Menschen und Menschengruppen. Der Soziologe Norbert Elias faßt sie als kollektive "Synthese-Leistungen" auf, als das Resultat der "sozialen Tätigkeit des Zeitbestimmens", durch welche Individuen und Gruppen kontinuierliche Ereignisreihen oder Geschehensabfolgen bilden und auf die Weise eigene Tätigkeiten miteinander und mit Abfolgen nichtmenschlicher Naturereignisse koordinieren resp. synchronisieren.[7]

Zeitkonstruktionen begrifflicher oder theoretischer Art beruhen immer auf lebensweltlichen Erfahrungen, die Menschen und Menschengruppen in konkreten Handlungskontexten machen, sowie auf den praktischen Zeitverhältnissen, die jene lebensweltlichen Erfahrungen implizieren oder die aus ihnen entwickelt werden können. In den konkreten Erfahrungskontexten der Lebenswelt haben die verschiedenen (praktische Zeitverhältnisse reflexiv verarbeitenden) Zeitkonstruktionen ihre ursprüngliche Plausibilität und Verständlichkeit. Wenn entwickelt, wirken Zeitkonstruktionen als Orientierungs- und Steuerungsmittel, indem sie weiteres Handeln als sinnvolle Tätigkeit ermöglichen und anleiten. Zeitkonstruktionen sind also Resultate von Reflexionsprozessen, in denen praktische Erfahrungen verarbeitet werden, und Orientierungsmedien, die weitere praktische Erfahrungen ermöglichen. Zeitkonstruktionen dienen der Selbstverständigung handelnder Subjekte, der Koordination und Steuerung menschlicher Praxis. Als Reflexionsresultate verarbeiten sie Erfahrungen, die in der Vergangenheit gemacht worden sind, und ermöglichen in der Gegenwart neue Erfahrungen und eine sinnvolle Praxis.

2. Die trinitätstheologische Geschichtskonstruktion des Joachim von Fiore

Hayden White unterscheidet in seiner sprachwissenschaftlich begründeten Metahistorie drei Grundformen der Geschichtsschreibung: die "Annalen", die "Chronik" und die eigentliche "Geschichte" oder "historische Erzählung". Die "Annalen" halten einfach fest, was der Fall gewesen ist. Da sie kein zentrales Thema haben, über das eine Geschichte erzählt werden könnte und das den einzelnen Ereignissen Wichtigkeit oder Signifikanz zuschreiben könnte, haben sie keinen Schluß. Sie schließen nicht ab. Sie hören einfach auf. Die Zeit, um die es in den Annalen geht, hat keine Höhepunkte oder Tiefpunkte. Sie ist parataktisch und endlos. Die "Chronik" hingegen stellt eine höhere Form der Konzeptualisierung der geschehenden Geschichte resp. eine überlegene historiographische Darstellunsgweise dar. Ihre Überlegenheit besteht in ihrem größeren Umfang, in der Organisation des verarbeiteten Materials nach z. B. Themenbereichen und Regierungszeiten sowie in ihrer größeren narrativen Kohärenz. Chroniken haben einen zentralen Gegenstand wie das Leben eines Individuums, einer Stadt oder einer Region, eine große Unternehmung (wie etwa einen Kreuzzug oder einen Krieg) oder eine Einrichtung (ein Kloster, eine Monarchie, ein Bistum). Chroniken, ähnlich wie die Annalen und anders als die Geschichtserzählungen, brechen einfach ab, statt abzuschließen. Ihnen fehlt die Geschlossenheit der von einem starken Sinnthema her durchkomponierten Geschichten.

Geschichtserzählungen, wie aus dieser Kontrastierung hervorgeht, haben einen "plot" (Hayden White), d.h. eine Sinnstruktur, die dem dargestellten Geschehen eine "geistige" Dimension verleiht und im Erzählzusammenhang der Geschichte eine Hierarchisierung von Wichtigerem oder weniger Wichtigem ermöglicht. In Geschichten geht es immer um Sinnhaftes, Signifikantes, von dem her die Kohärenz der Erzählung bewerkstelligt und eine Abgeschlossenheit des Erzählten historiographisch hergestellt wird. Die in einer Geschichte erzählten Ereignisse werden vom "plot" her bewertet und in ihrer Bedeutsamkeit klassifiziert.[8] Selbstverständlich sind die Bewertung und die Bedeutsamkeit der Ereignisse relative Größen: immer auf die jeweils gewählte Sinnstruktur hin bezogen und von den Erfahrungen und Interessen der Menschen und Gruppen abhängig, die die Geschichten erzählen und rezipieren. Geschichten haben immer ein Sinnzentrum. Deswegen verfahren sie selektiv, indem sie ihr Material gemäß gewissen Kriterien stark reflexiv verarbeiten, die als Kriterien nur in bestimmten Kontexten anerkannt sind und ihre Geltung haben.

Akzeptiert man diese von Hayden White vorgenommene grobe Klassifikation von Geschichtsdarstellungsformen,[9] so ist die heilstheologische Konstruktion des Abtes Joachim von Fiore eine "historische Erzählung" mit einer substantiellen "Sinnstruktur", die die ganze Geschichte der geschaffenen Welt zu erfassen vermag. Auf "typologischem" Wege entwickelt Joachim von Fiore seine sich als Heilsgeschichtskonstruktion präsentierende "historische Erzählung". Mit anderen Worten: mittels einer keineswegs neuen exegetischen Methode, nämlich der "typologischen" Methode, die angewendet wurde, um den verborgenen Sinn der heiligen Schrift zu erforschen, kommt Joachim von Fiore zu einer neuen Sicht des Heilsgeschehens, in der die zu seiner Zeit vorherrschende Auffassung radikal in Frage gestellt wird.

Die "Typologie" war eine konventionalisierte Bibelauslegungsmethode, durch die das von den Christen so bezeichnete Alte Testament in einem neuen Licht erscheinen konnte und die darin bestand, daß zwei Geschehnisse, Einrichtungen, Personen oder Dinge zusammengeschaut und auf die Weise zu einem sinnhaften Ereignispaar verbunden wurden. Das zustande gekommene Ereignispaar stellte einen Sinnzusammenhang dar, in dem die zwei involvierten Elemente sich wechselseitig beleuchteten. So zum Beispiel: die Opferung Isaaks durch Abraham im Alten und das Kreuzopfer Jesu Christi durch Gottvater im Neuen Testament, der Durchgang durch das Rote Meer im Alten und Jesu Taufe im Jordan im Neuen Testament, das Passahopfer im Alten und das Abendmahl im Neuen Testament, der Königin von Sabe Aufsuchen König Salomos im Alten und der Zug der heiligen drei Könige zum Jesuskind im Neuen Testament.[10] Im typologisch-allegorischen Auslegungsverfahren wurde das eine Element als der "Typos" (die "figura", "forma", Vorprägung oder Präfiguration) gedacht, der vom zweiten Element her, dem "Antitypos" (oder Ausprägung), interpretiert wurde und seinen Sinn bekam. Der "Typus" präfiguriert das, was später kommen sollte, d.h. den "Antitypus", der im Typus antizipiert wird und selbst dessen Steigerungsform darstellt.[11]

Die typologisch-allegorische Auslegungsmethode unterstellt, daß die Geschichte einen Sinn oder Zweck hat, und bemüht sich, diesen Sinn zu erforschen. Sie war zur Zeit Joachim von Fiores eine konventionalisierte Deutungs- und Lektürestrategie der heiligen Schrift, die Joachim von Fiore innovativ anwendet, indem er dazu übergeht, den Typus nicht mehr allein in der Vergangenheit zu suchen, sondern in der Gegenwart, und den Antitypus von der Zukunft zu erwarten, was zu einer realen Depotenzierung des Gegenwärtigen führen sollte. Der orthodoxen "evangelischen" Typologie, die die gegenwärtige Vollendung des in der Vergangenheit nur Angelegten verkündete, setzte Joachim von Fiore nun eine "apokalytische", "eschatologische" Typologie entgegen, die die bisherige Geschichte nicht mehr abschloß, sondern auf Zukünftiges, Kommendes hin radikal öffnete. Gegen die damals vorherrschende Geschichtssicht, nach der die Christen davon überzeugt waren, im letzten und vollkommensten Zeit- und Weltalter zu leben, verkündete Joachim von Fiore das Kommen eines neuen, dritten, noch vollkommeneren Weltreichs oder Weltalters, nämlich das Kommen des Weltzustandes ("status") des Heiligen Geistes, der "vollkommenen Gnade", der "Liebe" und der "Freiheit".[12] Mit Joachims eigenen Worten ausgedrückt:

> Wie wir schon in dieser Arbeit schrieben, überliefern uns die Losungen der heiligen Schrift schließlich drei Weltzustände: Den ersten, in dem wir unter Gesetz waren; den zweiten in der Gnade, den dritten, den wir in Bälde erwarten, in noch reicherer Gnade... Der erste Status war in der Wissenschaft, der zweite in der Macht der Weisheit, der dritte in der Vollkommenheit der Erkenntnis. Der erste in der Knechtschaft der Sklaven, der zweite in der Dienstbarkeit der Söhne, der dritte in der Freiheit. Der erste in Plagen, der zweite

> in der Aktion, der dritte in der Kontemplation. Der erste in der Furcht, der zweite im Glauben, der dritte in der Liebe. Der erste im Zustand der Knechte, der zweite der Freien, der dritte der Freunde... Der erste bringt Nesseln hervor, der zweite Rosen, der dritte Lilien. Der erste bringt Gras, der zweite Halme, der dritte Weizen. Der erste Wasser, der zweite Wein, der dritte Öl....[13]

Der von Joachim von Fiore mit heilstheologischen Begriffen und metaphorisch gebrauchten landwirtschaftlichen Ausdrücken beschriebene dritte Weltzustand ist ein status von Mönchen (nicht mehr von Verheirateten, wie der erste, oder von Klerikern, wie der zweite), in dem sich die Kontemplation und ein intensives völlig spiritualisiertes Leben entfalten werden: eine Zeit des Geistes, und nicht des Gesetzes, der Regelungen und des Buchstabens, eine Zeit der Freiheit und der Liebe.

In umständlichen Generationenberechnungen und Stammbäumen der heilsgeschichtlichen Zeiten mit ihren Geschlechterreihen untergliedert Joachim von Fiore seine drei Weltzustände, angeleitet vom Gedanken der Übereinstimmung oder Konkordanz beider Testamente. Die "concordia" von Altem und Neuem Testament bestimmt er folgendermaßen:

> Wir sagten, daß die Concordia in der Ähnlichkeit der gleichen Verhältnisse im Alten und Neuen Testament besteht: Gleich, sage ich, in Bezug auf die Zahl, nicht auf die Würdigkeit. Es stehen sich nämlich Person und Person, Ordnung und Ordnung, Krieg und Krieg in einer gewissen Gleichförmigkeit wechselseitig gegenüber. Zum Beispiel Abraham und Zacharias, Sara und Elisabeth, Isaak und Johannes der Täufer, Jesus als Mensch und Jakob....[14]

Die komplexen, zahlensymbolisch unterstützten Kalkulationen und Extrapolationen Joachims von Fiore dienen in der Tat seiner Bemühung der Herstellung einer Konkordanz zwischen der alten und der neuen Heilszeit. Diese Konkordanz besteht darauf, daß die Zeit von Christus bis zur Weltenruhe nach seiner Parusie die Zeit von Adam bis zu Jesus antitypisch spiegelt bzw. spiegeln soll. Kunstvoll vermag Joachim von Fiore dabei den Gedanken der (auf einer Zweiheit beruhenden) Konkordanz beider Testamente mit der die triadische Struktur der Trinitätslehre voll ausschöpfenden temporalen Konstruktion des Heilsgeschehens in drei Weltzuständen zu verbinden. Die letztendlich dadurch zustande kommende Disproportion und Asymmetrie zwischen beiden Testamenten wird qualitativ als zweiphasige Steigerung und Erfüllung des im Alten

Testament Angelegten bewertet. Dem Weltzustand des Alten Testamentes folgen zwei neue Weltzustände, die das Typische des ersten Weltzustandes antitypisch aufgreifen und vervollkommnen.

3. Zur Wirkungsgeschichte der trinitätstheologischen Geschichtskonstruktion Joachims von Fiore

Joachim von Fiore, der selbst zum zweiten Weltzustand gehörte, zog aus seiner eschatologischen Geschichtskonstruktion keine politisch revolutionären Schlüsse. Mit seiner Deutung des apokalyptischen Engels als des zur Erneuerung der christlichen Religion bestimmten "novus dux" verband er auch nicht die Absicht einer revolutionären Neugestaltung der bestehenden Einrichtungen und Sakramente. Joachim dachte streng theologisch; seine praktischen Reformen kamen einer radikalen Spiritualisierung des Mönchslebens gleich.[15] Dennoch stellten seine trinitätstheologische Geschichtskonstruktion und seine Prophezeiungen die traditionelle Autorität der etablierten Kirche und die zeitliche Ordnung des „saeculum de facto“ radikal in Frage, so daß sich die spätere politische Anwendung oder Verwertung seiner heilsgeschichtlichen Konstruktion als natürliche Folgerung aus seinem Denken ergab.

Alfons Rosenberg nennt verschiedene "Rezeptionsstationen" des Denkens Joachims von Fiore. Dabei handelt es sich um Bewegungen oder Autoren, auf die die temporale Konstruktion Joachims von Fiore entweder unmittelbar oder in einer subtil vermittelten Weise einen nachweisbaren Einfluß gehabt hat. Die von Alfons Rosenberg besonders hervorgehobenen Stationen sind: die Franziskaner-Spiritualen, Dante, Arnaldo von Villanova, Cola di Rienzo, Savonarola, Luther, Thomas Müntzer, Valentin Andreae, Besold, die Rosenkreuzer, Michael Maier und Robert Fludd, Campanella, Jakob Böhme, Lessing, Schelling, Franz von Baader, Rudolf Steiner, Kanndinsky, C.G. Jung und Leopold Ziegler.[16] Die Heterogenität des Denkens der von Alfons Rosenberg genannten Autoren und Bewegungen indiziert, wie vage der Begriff "Rezep-tion" häufig verwendet wird. Deswegen scheint es mir sinnvoll zu sein, im folgenden drei unterschiedliche Rezeptionsweisen des geschichtstheologischen Denkens Joachims von Fiore grundrißhaft vorzuführen, um anhand dieser drei verschiedenen Modi des Wirkens seines Werkes den Rezeptionsbegriff ein wenig zu präzisieren. Die hier ausgesuchten Rezeptionsmodi betreffen drei der von Alfons Rosenberg erwähnten Stationen: die Aufnahme des prophetischen Denkens Joachim von Fiores bei den Franziskaner-Spiritualen, Thomas Müntzers von den Visionen Joachims beeinflußtes revolutionäres Handeln und Friedrich Wilhelm Joseph Schellings "positive" Geschichtsphilosophie.

Handelt es sich bei der Rezeption der Geschichtskonstruktion und der Prophetien Joachim von Fiores bei den Franziskaner-Spiritualen um eine instrumentalisierende, weiterführende Anknüpfung seitens der Rezipierenden an das zu Rezipierende, so geht es in der revolutionären Praxis Thomas Müntzers um den Versuch einer Realisierung des von Joachim Prophezeiten, während die Schellingsche temporale Konstruktion eine die Konstruktion des Abtes nachahmende theoretische Figuration darstellt. "Instrumentalisierendes Anknüpfen", Versuch einer "Realisierung" und "analogisierendes Konstruieren" sind somit die drei Rezeptionsmodi oder Rezeptionsweisen des Denkens Joachim von Fiores, die nun kurz dargestellt werden sollen. Alle drei Rezeptionsmodi manifestieren und entfalten in je eigener Form den implizit subversiven Charakter des Rezipierten. Die Franziskaner-Spiritualen wenden das Rezipierte, an das sie bewußt anknüpfen und das sie weiterentwickeln, gegen Etabliertes und Geltendes an, durch das die ursprüngliche Botschaft des Evangeliums und das Ideal des apostolischen Lebens verraten wird. Thomas Müntzer wendet sich in seiner revolutionären Praxis der Realisierung des Rezipierten gegen soziale Verhältnisse, in denen Armut erzeugt und perpetuiert wird. Schelling schließlich opponiert mit seiner joachimanalogen "positiv"-philosophischen Geschichtskonstruktion gegen eine wirklichkeitsleere, geschichtslose, "logizistische" Philosophie, wie Hegel gemäß Schellings Interpretation sie praktiziert hat, sowie gegen eine eng rationalistische Kultur- und Lebensform.

Die zweite franziskanische Generation hatte in den Franziskanerorden, der neben dem Dominikanerorden als ein Bettelorden gegründet worden war, die prophetischen und geschichtstheologischen Lehren des Abtes Joachim von Fiore hineingetragen. Diese Lehren sollten dann eine wichtige Rolle in den großen Streitigkeiten spielen, die seit der Mitte des 13. Jahrhunderts den Franziskanerorden in zwei Parteien spalteten und in denen es um die Radikalität eines armen, apostolischen, evangelischen Lebens ging. Die eine Partei, die Partei der strengen Franziskaner-Spiritualen und Fraticellen in Frankreich und Italien, trat für ein solches radikales Leben ein und war nicht bereit, irgendwelche Kompromisse mit weltlichen Bedingungen oder Anpassungen an Etabliertes gutzuheißen. Überzeugt davon, daß der Franziskanerorden der von Joachim verheißene Geist-Orden der Zukunft sei, der alle irdischen Formen der bisherigen Welt, einschließlich der Organisationsform der Klerikerkirche, zu überwinden habe, setzten sich die Spiritualen für die Verbreitung der Lehren und Prophetien Joachims ein[17] und ergänzten die Prophetien des kalabresischen Abtes mit neuen, die sie selbst entsprechend den neuen Handlungskonstellationen und Auseinandersetzungsbedingungen nach dem Muster der Joachimschriften herstellten und zum Teil unter seinem Namen veröffentlichten.

Herbert Grundmann hat den Apokalypse-Kommentar des norddeutschen Minoriten Alexander und den "Liber de Flore", eine Schrift der Franziskaner-Spiritualen aus dem Anfang des 14. Jahrhunderts, untersucht und festgestellt, daß es sich bei beiden Schriften um pseudo-Joachimitische Schriften handelt, die lange als Werke von Joachim selbst betrachtet wurden, die aber aus dem Kreise der spiritualistischen Franziskaner stammen und das rasche und breite Wirken des eschatologisch-prophetischen Denkens Joachims in diesen Kreisen in hervorragender Weise dokumentieren.[18] Ernst Benz widmet in seiner grundlegenden Arbeit über die Ekklesiologie und die Geschichtstheologie der franziskanischen Reformation einige Kapitel der Wirkung joachimitischer Ideen im Franziskanerorden.[19] E. Benz' und H. Grundmanns Forschungen belegen, daß und wie im Franziskanerorden (bei den Spiritualen) Joachims Lehren gewirkt haben. Die spiritualistischen Franziskaner adaptierten und instrumentalisierten entsprechend ihren eigenen akuten Bedürfnissen Joachims Ideen. Indem sie an die trinitätstheologische Geschichtskonstruktion des Abtes hypoleptisch anknüpften, fanden sie ein Arsenal an Gedanken, Ideen und Vorstellungen, die sie für ihren eigenen Ordenskampf verwenden konnten. Sie nahmen die eschatologisch-prophetischen Ideen und Gedanken Joachims auf, und sie entwickelten sie weiter. Zweifelsohne stellen die Kreise der spiritualistischen Franziskaner dasjenige Rezeptionsmilieu dar, innerhalb dessen Joachims Ideen am wirksamsten gewesen sind.[20]

Anders als bei den spiritualistischen Franziskanern geht es bei Thomas Müntzer und bei Friedrich Wilhelm Joseph Schelling um eine Rezeption in einem weiten und viel vermittelteren Sinne. Das Soziale und das Politische spielen im Denken Joachim von Fiores selbst keine zentrale, explizite Rolle, konzentriert sich doch die Aufmerksamkeit des Abtes auf den christlichen Weg der "perfectio" und auf das Mönchsideal. Dennoch implizieren seine apokalyptisch-eschatologischen Anschauungen Soziales und Politisches, was Thomas Müntzer keineswegs entgangen ist. In seinem radikalen, über Luther hinausgehenden Kampf gegen das etablierte, korrumpierte Kirchensystem agierte Müntzer als eschatologischer Prophet, der konsequent auf die Armen setzte, da diese aufgrund ihrer unterstellten größeren Distanz gegenüber den Versuchungen der "avaritia" und "luxuria" potentielle Auserwählte seien. In seinen Predigten relativierte Müntzer die Gegenwart und eröffnete prospektiv-visionär mittels der tradierten apokalyptischen Texte (insbesondere des Buches Daniel) einen Zukunftsraum, in dem Ausbeutung und luxuriöser Reichtum auf Kosten der Armen nicht existieren würden.

Revolutionär handelnd setzte sich Müntzer für die Realisierung eines solchen ganz egalitären Zukunftsraums als Reich der Auserwählten ein. Mit anderen Worten: Was Müntzer treibt, ist die Realisierung der Idee des Reiches Gottes. Seine konkrete, zu implementierende Botschaft lautet: Die "armen Laien und Bauern" sollen Träger der staatlichen Gewalt und damit Repräsentanten der göttlichen Gerechtigkeit auf Erden sein. Jener von Alfons Rosenberg zitierte Satz Müntzers "Bei mir ist das Zeugnis Abbatis Joachim groß" indiziert, daß Müntzer in seinen Predigten explizit an den Abt Joachim erinnert hat, resp. dessen Prophetie für sein eigenes Handeln eine legitimierende und inspirierende Rolle gehabt hat.[21]

In seinen Vorlesungen über die "Philosophie der Offenbarung" will der Philosoph Friedrich Wilhelm Joseph Schelling das Christentum als "geschichtliche Tat-Sache" begreifen und erklären. Philosophie faßt der späte Schelling als "die Vernunftwissenschaft" des "Seienden" auf, dessen also, was wirklich, "tat-sächlich" und geschichtlich ist. Philosophie ist für Schelling demnach keine Gedankenkonstruktion, keine geschichtslose Logik, sondern der Begriff und die Erklärung des Wirklichen.[22] Ist das Christentum höchst wirklich, ja die höchste Wirklichkeit und geschichtliche Tatsache, die "Alles" zu sein beansprucht,[23] so muß die Vernunftwissenschaft des Wirklichen wissenschaftliche Erklärung des Christentums, "positive" Philosophie der christlichen Offenbarung werden. In ihr wird dann das Christentum in seiner geschichtlichen Wirklichkeit erfaßt, nämlich als "tat-sächliche", wirkliche "Offenbarung", die das menschliche Leben real bestimmt. Zur geschichtlichen Realität des Christentums als Offenbarung gehört die Kirche, deren Geschichte Schelling philosophisch rekonstruiert. Diese philosophische Rekonstruktion der Geschichte der christlichen Kirche knüpft an Joachims triadische Geschichtstheologie explizit an.[24] Schelling unterscheidet analog zu Joachim von Fiore drei verschiedene Perioden: die petrinische, die paulinische und die johanneische Periode. Alle drei sind notwendige Phasen in der geschichtlichen Entfaltung des wahren Christentums. Ist die petrinische Periode der "Anfang" und der "Grund", so stellt die paulinische Periode die "Bewegung" und die "Freiheit" des christlichen Lebens dar. Die johanneische Periode ist dann die Epoche des "Geistes", der "wahren Allgemeinheit", der "Vollkommenheit" und der "höheren" Gestalt des Christentums. Dem "Bleibenden", "Substantiellen" und "Gesetzlichen" der petrinischen Epoche folgt demnach die lebendige Freiheit der paulinischen Epoche, die im Protestantismus realisiert wird. Dieser dynamischen Epoche wird die wahre Kirche des Johannes folgen, welche eine "vollkommene Verschmelzung des Christentums mit der allgemeinen Wissenschaft und Erkenntnis" mit sich bringen wird. Prägnant heißt es im von Schelling selbst redigierten Text:

"Petrus ist der Gesetzgeber, der Grundlegende, das Stabile. Paulus brach hervor als ein Feuer, ist der Elias, repräsentiert das Prinzip der Entwicklung, der Bewegung, der Freiheit in der christlichen Kirche. Johannes ist der Apostel der Zukunft."[25] Um diese Zukunft geht es Schelling. In ihr wird die Philosophie eine substantielle Tätigkeit sein, die das Wirkliche, Geschichtliche und Relevante erfaßt. Auf diese Zukunft kommt es auch an: eine Zukunft der Versöhnung, der Vollkommenheit, der Lebendigkeit, des "sanft wehenden, verklärten Geistes" und der Fülle des Lebens.

Schelling, genauso wie die franziskanischen Spiritualen und Thomas Müntzer, unzufrieden mit der Gegenwart, mit dem Ist-Zustand einer steril gewordenen "logizistischen" Philosophie und einer zerrissenen, eng-rationalistischen Kulturwelt, öffnet den Gegenwartsraum radikal auf Zukunft hin. Ohne die Inspiration und das Orientierungsmodell des Abtes Joachim von Fiore wäre die konkrete Gestalt der Schellingschen radikal prospektiven Geschichtskonstruktion nicht möglich gewesen.

Universität St. Gallen/Freie Universität Brüssel (ULB)

Anmerkungen

1. Wenn man mit den klassischen Geschichtphilosophen "Geschichte" als das auffaßt, was Menschen handelnd "mit Willen und Bewußtsein" machen und gestalten können, dann ist die Rede von mittelalterlichen Geschichtskonzeptionen oder Geschichtsanschauungen problematisch, da es im Mittelalter "die" Geschichte gar nicht gibt, die erst in der Aufklärung als mit Vernunft zu gestaltende Größe the-matisiert wird. Dennoch ist es gerechtfertigt, in einem weiten Sinne von Geschichte zu sprechen und den Begriff auf jene "Veranstaltung" anzuwenden, welche gemäß den heilsökonomischen theologischen Traktaten Gott zur Rettung des gefallenen sündigen Menschen stattfinden läßt.

2. Selbstverständlich ist der Begriff des lateinischen Averroismus eine nicht ganz präzise "etikettierende" Bezeichnung für die Entstehungs- und Wirkungs-geschichte einiger Ideen, die im 13. Jahrhundert an der Pariser Universität von den Philosophen entwickelt und propagiert wurden und die sich nicht mit verschiedenen Inhalten des christlichen Glaubens versöhnen ließen, was zu offiziellen Ver-urteilungen (1270 und 1277) seitens der kirchlichen Behörden führte. Dennoch ist der Begriff berechtigt, ermöglicht er doch, einen wichtigen Denkzusammenhang zu bezeichnen, der als lebendige philosophische Richtung viele Diskussionen und Kontroversen angeregt und markiert hat. Vgl. Friedrich Niewöhner, Loris Sturlese (Hrsg.), *Averroismus im Mittelalter und in der Renaissance* (Zürich: Spur Verlag, 1994) sowie Albert Zimmermann (Hrsg.), *Albertus Magnus und der lateinische Averroismus,* in Gerbert Meyer, Albert Zimmermann (Hrsg.), *Albertus Magnus - Doctor universalis 1280/1980* (Mainz: Matthias-Grünewald-Verlag, 1980), 465-93. Zu den wichtigsten Lehren des

Hauptrepräsentanten des Averroismus, Siger von Brabant, vgl. Fernand Van Steenberghen, *Maître Siger de Brabant* (Löwen: Louvain Publications Universitaires,1977), 258ff.

3. Vgl. Wolf-Ulrich Klünker und Bruno Sandkühler, *Menschliche Seele und kosmischer Geist. Siger von Brabant in der Auseinandersetzung mit Thomas von Aquin. Mit einer Übersetzung der Schrift Sigers De anima intellectiva (Über die Geistseele)* (Stuttgart: Verlag Freies Geistesleben, 1988) sowie die Einleitung von Alain de Libera, in Thomas d'Aquin, *L'unité de l'intellect contre les averroïstes* (Paris: Flammarion, 1994), 9-73.

4. Zur Wirkungsgeschichte dieser naturalisierenden Sicht des Weltgeschehens vgl. Herbert Grundmann, *Ausgewählte Aufsätze*, Teil 2: Joachim von Fiore (Stuttgart: Hirsemannn, 1977), 217ff.

5. Die verschiedenen Phasen oder Stationen dieser teilweise subversiven Wirkungsgeschichte nennt Alfons Rosenberg in seiner Einleitung zu der Anthologie der für die Heilsgeschichtskonzeption Joachims von Fiore einschlägigen Texte des Abtes von Calabrien. Vgl. Joachim von Fiore, *Das Zeitalter des Heiligen Geistes*, hrsg. und eingel. von Alfons Rosenberg (Bietigheim: Turm Verlag, 1977), 50ff. Ich habe mich entschieden, die Werke des Abtes Joachim von Fiore nach dieser Ausgabe zu zitieren, selbst wenn sie textkritischen Ansprüchen nicht ganz genügt. Ihre all-gemeine Zugänglichkeit sowie die Tatsache, daß sie alle für die Geschichtstheologie Joachims von Fiore einschlägigen Texte enthält, mögen eine solche Entscheidung rechtfertigen.

6. Vgl. Edmund Husserl, *Gesammelte Schriften*, hrsg. von Elisabeth Ströker, Bd. 8 (Hamburg: Felix Meiner, 1992),132. Zum von Husserl wissenschafts- und objektivitätskritisch verwendeten Begriff der Lebenswelt vgl. Thomas Gil, "Soziale Interaktion, Verständigungshandeln und wissenschaftliche Begriffsbildung. Zur Sozialpragmatik naturphilosophischer Grundbegriffe und Diskurse," in Holger Burckhart (Hrsg.), *Diskurs über Sprache* (Würzburg: Könighausen & Neumann, 1994), 69-82.

7. Vgl. Norbert Elias, *Über die Zeit* (Frankfurt a.M.: Suhrkamp, 1988), 1f., 45 und 100.

8. Vgl. Hayden White, *Die Bedeutung der Form. Erzählstrukturen in der Geschichtsschreibung* (Frankfurt a.M.: Fischer Verlag, 1990), 19, 20ff., 28f. und 37. Vgl. auch Hayden White, *Auch Klio dichtet oder Die Fiktion des Faktischen. Studien zur Tropologie des historischen Diskurses* (Stuttgart: Klett-Cotta, 1991).

9. In seiner "Geschichtsschreibung im Mittelalter" unterscheidet Herbert Grundmann folgende Gattungen der Geschichtsüberlieferung und Geschichtsdarstellung: die volkssprachliche Geschichtsdichtung (Lieder und Sagen), die Volksgeschichte (*origo gentis*), Weltchroniken, Annalen, Vitae, Gesta, Landes und Stadt-chroniken sowie die lateinische Volksdichtung. Vgl. Herbert Grundmann, *Geschichtsschreibung im Mittelalter* (Göttingen: Vandenhoeck & Ruprecht, 1987), 4.

10. Zur bibelexegetischen Methode der "Typologie" vgl. Volker Bohn (Hrsg.), *Typologie. Internationale Beiträge zur Poetik* (Frankfurt a.M.: Suhrkamp, 1988), insbes. die Beiträge von Friedrich Ohly ("Typologie als Denkform

der Geschichts-betrachtung") und Northrop Frye ("Typologie als Denkweise und rhetorische Figur"), 22-63 und 64-96.

11. Die typologische Interpretation stellt eine Strategie der Depotenzierung und Enteignung des sogenannten Alten Testaments dar, wurde dieses nur im Lichte des Neuen Testaments und von diesem her gedeutet. Jüdische Gelehrte protestierten immer wieder gegen diese "Enteignungsstrategie", die Christus und sein Kreuz überall in den einzelnen Schriften des Alten Testaments präfiguriert sah.

12. Die drei Hauptschriften Joachims sind die "Concordia novi et veteris testamenti", die "Expositio in Apocalipsim" und das "Psalterium decem chordarum". Für die Lehre von den drei Zeitaltern sind die Hauptquellen die Einleitung und Kapitel V der "Erklärung der Apokalypse" sowie die "Konkordanz", Buch V, Kapitel LXXXIV. Alle einschlägigen Texte für die Geschichtstheorie Joachim von Fiores liegen vor, in: Joachims von Fiore, *Das Reich des Heiligen Geistes*, hrsg. und eingel. von Alfons Rosenberg (Bietigheim: Turm, 1977). Zur Geschichtstheologie Joachim von Fiores im allgemeinen vgl. Herbert Grundmann, *Studien über Joachim von Floris* (Leipzig: B.G. Teubner, 1927), 56ff.

13. Joachim von Fiore, *Das Reich des Heiligen Geistes*, 82f. Die vielen land-wirtschaftlichen Metaphern und Beispiele, die in Joachims Schriften enthalten sind, haben unter Joachim-Gelehrten die Vermutung aufkommen lassen, Joachim sei Bauer gewesen und habe seinen Acker bestellt, bevor er ins Kloster eintrat. Das landwirtschaftliche Erfahrungswissen, das als Basis für seine Redeweise fungiert, bestätige, so heißt es, eine solche Hypothese. Vgl. Herbert Grundmann, *Ausgewählte Aufsätze*, Teil 2: Joachim von Fiore, 89.

14. Joachim von Fiore, *Das Reich des Heiligen Geistes*, 79f.

15. In diesem Sinne sind auch seine Ordensgründung und seine verschiedenen Strukturierungsvorschläge des Klosterlebens zu verstehen. Vgl. die Einleitung von Alfons Rosenberg, in Joachim von Fiore, *Das Reich des Heiligen Geistes*, 12ff.; die verschiedenen Texte zum Neuen Orden und Klosterplan (in: Joachim von Fiore, *Das Reich des Heiligen Geistes*, 133ff. und 147) sowie Karl Löwith, *Weltgeschichte und Heilsgeschehen. Die theologischen Voraussetzungen der Geschichtsphilosophie* (Stuttgart: Kohlhammer, 1953), 141ff.

16. Vgl. Joachim von Fiore, *Das Reich des Heiligen Geistes*, 53ff.

17. So veröffentlichte beispielsweise in Paris im Jahre 1254 der Franziskaner Gherardo di Borgo San Donnino die drei Hauptwerke Joachims mit einer Einführung dazu als "Evangelium aeternum". Diese zusammengestellte und bewußt per-spektivierte Schrift wurde von den Pariser Theologie-Professoren als eine Provokation empfunden und sollte später vom Papst (Alexander IV.) verurteilt werden. Thomas von Aquin nahm auch Stellung, indem er in drei *Quaestiones* seiner *Summa theologiae* (Prima secundae, quaestiones 106-108) klar darlegte, daß das Alte Testament und sein Gesetz (die "vetus lex") in und durch Christus erfüllt und aufgehoben seien, dessen "nova lex" ("lex evangelica, quae dicitur lex nova") bis zu seiner Wiederkunft am Ende der Zeiten unabänderlich gelte. Deswegen sei es "vanitas", nach der Zeit des "neuen Gesetzes", die ja bis zum Weltende dauern werde, ein anderes "tempus Spiritus Sancti" zu erwarten. Zum "Evangelium aeternum" vgl. Herbert Grundmann, *Ausgewählte Schriften*, Teil 2: Joachim von Fiore, 225 und 408. Zur Auseinandersetzung

Thomas von Aquins mit den Lehren Joachims vgl. Ernst Benz, *Ecclesia Spiritualis. Kirchenidee und Geschichtstheologie der franziskanischen Reformation* (Stuttgart: Kohlhammer, 1934), 181ff. sowie Max Seckler, *Das Heil in der Geschichte. Geschichtstheologisches Denken bei Thomas von Aquin* (München: Kösel, 1964), 189ff.

18. Vgl. Herbert Grundmann, *Ausgewählte Aufsätze*, Teil 2: Joachim von Fiore, 58ff. und 101ff.

19. Vgl. Ernst Benz, *Ecclesia Spiritualis*, 3ff., 175ff. und 244ff.

20. Vgl. Arno Borst, *Barbaren, Ketzer und Artisten: Welten des Mittelalters* (München: Piper, 1988), 232ff., und Norman Cohn, *The Pursuit of the Millennium. Revolutionary Millenarians and Mystical Anarchists of the Middle Ages* (London: Paladin, 1957), 40, 110, 158 und 224.

21. Vgl. die Einleitung A. Rosenbergs, in Joachim von Fiore, *Das Reich des Heiligen Geistes*, 57. Dort ist die Rede davon, daß Müntzer sich durch Joachims Prophetie "legitimiert" und "befeuert" gefühlt hätte. Es gibt eine kritische Gesamt-ausgabe der Schriften und Briefe Müntzers: Thomas Müntzer, *Schriften und Briefe. Kritische Gesamtausgabe*, unter Mitarbeit von Paul Kirn, hrsg. von Günther Franz (Quellen und Forschungen zur Reformationsgeschichte, Bd. XXXIII), (Gütersloh: Verlagshaus Gerd Mohn, 1968). Zu den Inhalten und der Richtung der Müntzerschen Predigten und Theologie vgl. Ernst Bloch, *Thomas Müntzer als Theologe der Revolution* (Berlin: Aufbau Verlag, 1960), sowie Siegfried Bräuer, Helmar Junghans (Hrsg.), *Der Theologe Thomas Müntzer. Untersuchungen zu seiner Entwicklung und Lehre* (Göttingen: Vandenhoeck & Ruprecht, 1989).

22. Schelling wirft Hegel vor, die philosophische Begriffsarbeit so betrieben zu haben, daß aus der höchsten Wirklichkeitswissenschaft, nämlich der Philosophie, eine wirklichkeitsleere "Logik" geworden sei, der es unendlich an Sein mangele. Vgl. Manfred Frank, *Der unendliche Mangel an Sein. Schellings Hegelkritik und die Anfänge der Marxschen Dialektik* (Frankfurt a.M.: Suhrkamp, 1975) sowie Axel Hutter, *Geschichtliche Vernunft Die Weiterführung der Kantischen Vernunftkritik in der Spätphilosophie Schellings* (Frankfurt a.M.: Suhrkamp, 1996).

23. Vgl. Friedrich Wilhelm Joseph Schelling, *Philosophie der Offenbarung*, 1841/42, hrsg. und eingel. von Manfred Frank (Frankfurt a.M.: Suhrkamp, 1977), 79f. und 309f.

24. Ibid., 315.

25. Ibid., 317.

Astrolabes and the Construction of Time in the Late Middle Ages

Edgar Laird

The measurement of time in the late Middle Ages was often referred to the rotation of the celestial sphere -- the sphere, that is, in which all celestial positions are located, called the "primum mobile" in the old cosmology. The kind of astrolabe most in use during the same period, the planispheric astrolabe, was so constructed that it could simulate the rotation of the celestial sphere by turning a star-map called the "rete" over a disc marked with coordinate lines for the observer's horizon.[1] Thus the motion of the sphere, and hence time, could be divided and measured. In practice, therefore, the problem of finding exact time was virtually identical with that of determining exact locations of known points in the celestial sphere, for example, the position of a fixed star or the Sun's ecliptic longitude (i.e., its place in the zodiac). An astrolabist located such points in relation to coordinates fixed by his meridian and horizon, and in so doing he established local time and corrected the errors of clocks and other chronometric devices. For someone actually measuring time, questions of an Augustinian sort about the nature and subjective construction of time need not have arisen.[2]

Techniques employed in measuring time could, nevertheless, imply theoretical assumptions, as is already apparent in the practice of referring time to the *primum mobile*, a practice that invokes the much-discussed Aristotelian principle that "Time is the measure of motion in the primum mobile."[3] The purpose of the present essay is to survey various uses of the astrolabe in various contexts and to draw inferences about astrolabes and the construction of time in the late Middle Ages. The survey draws mainly from treatises on the composition and usage of the astrolabe. For the period under consideration, the most important ones are as follow:

1) *Compositio et operatio astrolabii*, attributed in the Middle Ages to the eight-century Arabic writer Messahalla (Masha'allah). A Latin version assembled and re-edited in the thirteenth century became a standard text for teaching for the next few centuries.[4]

2) Pèlerin de Prusse, *Practique de astralabe*, written in 1362, in French, for Charles, later Charles V of France.[5]
3) Geoffrey Chaucer, *Treatise on the Astrolabe*, written in 1391 or shortly thereafter, in English, ostensibly for a ten-year-old boy, "Lowys."[6]
4) Jean Fusoris, *Composition et usage de l'astrolabe*, written between 1407 and 1412, in French, and dedicated to Peter of Navarre, Count of Mortrain. It is probably identical with the astrolabe treatise Fusoris is said to have presented to Robert Courtenay, Bishop of Norwich, and Henry V of England. It was translated into Italian in the late fifteenth century.[7]

In what follows, earlier and later treatises will also be mentioned occasionally, for purposes of comparison and contrast.

The astrolabe, from its origin in Greek antiquity through its Islamic development and its early career in the West, evolved in the direction of greater accuracy and precision; but from the thirteenth century through the fifteenth it was very stable, having achieved a limit of technical perfection that left little need or room for change. The later medieval treatises consistently show that in a characteristic operation the astrolabist begins by determining the Sun's ecliptic longitude (which is tied to the day of the year) from a calendar scale on the back of the instrument. Next he uses the astrolabe's alidade or sighting rule to measure the Sun's altitude above the horizon (tied to the time of day). He then arranges the parts of the astrolabe that represent the celestial sphere and the coordinates of his horizon so that the analogue of the Sun's longitude coincides with that of its altitude. From this arrangement he can read off the time and the "ascendant" (i.e., the degree of the zodiac on the eastern horizon), from which an astrological horoscope can be established. Thus the astrolabe furnishes basic data for its two principal offices: time-telling and astrology.

The procedures just sketched are, quite reasonably, given at the beginnings of the "usage" sections of the astrolabe treatises. "Messahalla," as we shall designate that treatise, gives directions by which "to know the exact time and also the ascendant" ("scire certitudinem hore et etiam ascendentis").[8] Pèlerin says he is explaining how "to know the time of day and night and the sign and degree of the ascendant" ("savoir quelle heure il est des jours et de nuit et le signe et degré ascendent"),[9] thus making clear what unit of time he is noting and also that the ascendant is a degree of a zodiacal sign. Chaucer's statement of intention is still more explicit:

> To knowe every tyme of the day by light of the sonne; and every tyme of the nyght by the sterres fixe; and eke to knowe by nyght or by day the degre of eny signe that ascendith on the est orizonte, which that is clepid comounly the ascendent, or ellis horoscopum.[10]

He thus indicates that finding the time and ascendant at night, when the Sun in invisible, can be achieved by noting the location of a star. (As already stated, any point in the celestial sphere will do as long as it is represented on the astrolabe, for it is by the celestial sphere that time is told.) Finding the ascendant degree or "horoscopum" allows one to begin framing a horoscope chart. Fusoris's directions are "to know the time ... and the four principal houses of the sky" ("savoir les heures ... et les 4 maisons principles du ciel").[11] Astrologers recognized twelve astrologically significant segments of sky called "houses." Of the four principal ones the first or most principal was the ascendant house, which Fusoris says begins at the degree ascending on the east horizon ("degré ... ascendent c'est a dire en l'orizon orient"). Marking out the four chief houses instead of just one is another step toward casting a horoscope.

Another way to state the case to which these writers refer is to say that the astrolabe represents three parameters relating to the celestial sphere: the day of the year, the altitude of the Sun (or that of a known star), and the hour of the day. Given any two, one can find the third. A single setting of the astrolabe represents all three, and a single operation is required to establish the hour, the day, and the framework of a horoscope for that hour and day. Astrolabe treatises tend to keep the time and the horoscope separate and indeed to suppress astrological implications in favor of the time-reckoning aspects of celestial motion. Astrology has interesting bearings on the construction of time, and these must be discussed. But following medieval practice we will make a separation and discuss first those features and uses of the astrolabe that have more purely to do with time.

It appears to have been the time-telling capabilities of the astrolabe that first made it welcome in the West. The oldest extant Western text on the astrolabe (late tenth century) opens with the statement that a knowledge of the astrolabe and its parts is necessary for anyone who wishes to know the certain hours of night and day or the smaller divisions of time ("scire certas horas noctium et dierum vel punctos et momenta").[12] This treatise was a product of the Benedictine monastery of Santa Maria de Ripoll in Catalonia and was influenced by and partly translated from Islamic work on the astrolabe. Its own influence quickly spread in the eleventh century to other monastic houses -- at Fleury, Reichenau, and Malvern, that we know of[13] -- and indeed the early history of the astrolabe in the West is to be found among the monks, who almost uniquely had the motive and skill for telling time by precise astronomical means. In Islam astrolabes had been adapted for calculating times of prayer. Among early Western innovations was the development of techniques for establishing the beginning of the Christian year,[14] and from the first the astrolabe was adapted to the fixing of canonical hours. For example tierce and nonce, the third and ninth hours, could not conveniently be reckoned at a uniform 9 a.m. and 3 a.m., but they could easily be reckoned with an astrolabe at half-way through the first part of the day and half-way through the second.[15] Beyond measuring hours of the day, the monasteries applied the astrolabe to problems of when, exactly, the Sun rises and sets; when the Moon passes through its phases; how these motions relate to the days of the week, the lunar month (and hence the reckoning of Easter and other movable feasts), and the solar year; and in general how all of these organize the ecclesiastical and secular calendars.[16] The subtle mathematical calculations that underlie such problems are mechanized, so to speak, by the astrolabe.

Its usefulness and the comparative ease of its use did not go unnoticed by a wider public. By the twelfth century the astrolabe had become a popular symbol of astronomical learning, and it was known at least by repute to all who saw it depicted in cathedral sculpture and in miniatures.[17] By the fourteenth and fifteenth centuries, treatises in Latin and vernaculars were teaching varied audiences how to use the astrolabe. The treatises and the instrument itself provide evidence of the interests and needs of the societies that employed them.

For example, the treatises deal with two kinds of hours, "unequal" and "equal." The unequal hours are appropriate to a mainly rural society in which activities depend upon daylight and night-work; they were, however, felt to be vaguely unnatural or disorderly.[18] These unequal hours are obtained by dividing daylight into twelve hours and night into twelve. As the length of days and nights vary according to seasons and one's geographical latitude, so do the lenghts of unequal hours of local time. When hours of this kind were measured by candles notched to mark them, the spaces between notches had to be varied, by changing and roughly estimated amounts, from season to season and from place to place. With an astrolabe, division into unequal hours for various seasons and places was easy and precise.[19] For scholarly astronomers, as well as for travellers and university-trained physicians, it was preferable to use the more universally applicable equal hours, obtained by dividing the period of one revolution of the celestial sphere into twenty-four equal parts. Astronomers, who as experts on celestial motion were also experts on time, were often required to convert from their system of equal hours into the system of unequal hours that their patrons and the public held to be more relevant to their needs. It is not surprising therefore that the astrolabe is devised so that at the turn of a rule the conversion can be accomplished in either direction.[20]

It is possible that repeated acts of conversion, so regular that they could be mechanized, may have insinuated that the division of time is in some ways quite arbitrary. One may divide this way or that, and in fact even the mathematical tools that the astrolabe embodies could be seen to have about them an element of arbitrariness and convention, for as Nicole Oresme pointed out in the later fourteenth century, a celestial circle or any circle at all need not have been divided into 360°: it could have been divided, if one had wished, into 480° or any other number. It is all a matter of will and arithmetic.[21] By another ancient convention an equal hour is the period of 1/24 of a rotation of the celestial sphere, hence 15°. Accordingly, Messahalla says that in order to measure a duration one should take the number of degrees through which the celestial sphere has turned and divide by fifteen.[22] Pèlerin, perhaps simplifying the arithmetic for a courtly audience, says to move the astrolabe's rule across the arc of rotation and count the number of times it passes fifteen degrees.[23] Neither procedure is necessary with the astrolabes Chaucer and Fusoris describe, for they are marked with hour-divisions alongside the number-divisions representing degrees of arc.[24] In short, the conventions relating degrees to hours, already quite firm, have been literally inscribed in the astrolabe.

Within the conventions on which the astrolabe is constructed a certain amount of freedom is possible, as may be seen in the use of the instrument as a kind of intellectual toy. The treatises on usage do not limit the number of operations that can be performed. They might explain any number, from twenty (Pèlerin) up to fifty or so (*Compositio brevis*, 14c.),[25] and the number of usages in Messahalla and Chaucer varies from manuscript to manuscript, with some additions being evidently due to scribes or compilers.[26] The assumption seems to be, as Pèlerin makes explicit, that after one has mastered a few fundamental techniques, he can find or freely invent others.[27] Some described in the treatises are redundant, offering additional ways to solve a problem -- finding the length of an unequal hour, say, or the period from dawn to sunrise -- that have already been solved otherwise. These serve no useful purpose but are, as Chaucer says, "for a maner curiosite."[28]

Yet, the limits on freedom to play are finally quite strict, fixed by the objective something that Aristotle says underlies time.[29] That the objective something is celestial motion is implicit in the astrolabe. The mathematical model engraved as lines and points on an astrolabe is an abstraction from the celestial spheres, a circumstance to which may be applied the Aristotelian dictum that abstraction is not a lie.[30] The truth of the abstract model is bound by the truth of the original. Elsewhere such questions might arise for debate as whether time is perspectival or extrinsic and absolute, but astrolabe treatises remain unpolemical users' manuals for an instrument constructed on stable principles. The treatises of Messahalla and Pèlerin are strictly technical, and even Chaucer, a poet with large human and philosophical interests, produced an astrolabe treatise which, like that of the artisan and commercial entrepreneur Fusoris, excludes all speculative problems, all questions of the definition of space and time, all the controversies that agitated scholastic physics in the late Middle Ages.[31] These writers silently assent to the theoretical givens of their instrument. They introduce no potter's wheel, no motion or conception in the human soul, no exemplars in the Divine Mind in which to ground time. Their mental exercises with the astrolabes are wholly within the rules of its celestio-mathematical model, and they claim for it a high degree of exactness.

If a laborer in the field wanted to know when to eat his midday meal, it was enough for him to see when he could comfortably step on the shadow of his head. Exact time, on the other hand, was calculated by abstract mathematical astronomy, whose instrument was the astrolabe. Astronomical time was mathematical time.[32] It was the most certain time because mathematical knowledge was seen as more certain than any other human knowledge.[33] When in the Middle Ages horary information is given with precision, chances are that it results from calculations and not from clepsydra, candle, or clock.[34] When we read of "*certain* hours" in a medieval text, these are not, as has been asserted, a fourteenth-century bourgeois innovation displacing "the *uncertain* clerical hours of the church bells."[35] They are the certain hours expressed by the phrase *certas horas*, a technical term used in monastic and scholarly astrolabe treatises to denote mathematically exact times calculated for ecclesiastical or scholarly use.[36] These hours could, of course, be transferred to more public uses in government and commerce. Indeed there is no doubt that they were. Their history, however, is bound up with the history of the astrolabe, which is, by one repeated definition, the instrument by which the certain hours ("certas horas") are known.[37]

The treatises of Pèlerin and Fusoris, written in the fourteenth and fifteenth centuries respectively, are addressed to princes and magnates. Even as early as the twelfth century, an astrolabe treatise by Adelard of Bath was addressed to Henry Plantagenet, later Henry II of England.[38] Chaucer's *Astrolabe*, written near the end of the fourteenth century, claims to address a ten-year-old boy, but the claim may be in part a pretense by which Chaucer could write plainly and simply for a wider audience, including courtly personages, without seeming uncourtly or condescending. Clearly he anticipated an audience of more than one,[39] and fifteenth-century manuscripts of his work suggest a varied readership.[40] In any case, the audiences of later astrolabe treatises differ from those of the early ones written for monastic use. The later treatises suggest that the astrolabe was useful in some unspecified way or ways to those responsible for public governance.

What might its uses have been? Certainly, for purposes of planning military campaigns and anticipating times of agricultural plenty and scarcity, it was desirable to know what has been aptly termed "meteorological time."[41] Late medieval attempts at a precision beyond available means led to the development of a professional discipline modern historians call "astrometeorology."[42] Almost all of the quite numerous treatises in this discipline appear to have been written by and for professionals, and they became increasingly professionalized through the fourteenth and fifteenth centuries. They are not addressed to princes and great landowners. A second possible use for the astrolabe derives from the suggestions that the study of celestial order has ethical value for the ordering of a great man's character. The idea had been put forward by Ptolemy, and Christine de Pizan reproduces it in her early-fifteenth-century educational biography of Charles V, as having been a motive for Charles's star-study and as an encouragement to star-study by his successors.[43] One of the later astrolabe treatises, that of Fusoris, suggests that an exalted person such as Peter of Navarre does well to study the astrolabe because its appeal is to the spiritual part of a man, through the goods of virtues and knowledge, and not to the body, which is only ordure and putrefaction.[44] This possible use of the astrolabe is not, however, related directly or simply to time.

A third possible use, finding the ascendant and framing a horoscope, has already been mentioned as having a direct bearing on the construction of time. Warnings that star-study might actually be *in*appropriate for princes and magnates hint at a fear that astrolabes might be used for judicial astrology, to the detriment of public governance.[45] Two parts of astrology, "nativities" and "elections," relate to time in two different ways. There are, as Pèlerin says, "two manners" of arranging a setting of the astrolabe.[46] One is to find the time and set the instrument to represent the heavens at that time. This manner is suited to nativities, when the astrologer wishes to know the state of the heavens at the time of a birth. The other is to set the astrolabe in representation of celestial conditions the astrologer is interested in and to determine when those conditions will be met, for the astrolabe permits calculations, as

Pèlerin explains, for any time "quelconques journee present ou autre."[47] This manner is suited to elections, by which one determines the best time to perform an action.

For our purposes, nativities can be viewed chiefly as providing yet one more motive for precision in time-telling and can be treated briefly.[48] In order to account for such things as differences between twins, in character and destiny, astrologers had to posit atomically precise temporal determinations.[49] Their horoscopes became mathematical abstractions, with the minutest part of a degree marking the minutest instant of time. One early astrolabe treatise demonstrates that as number is the element of arithmetic and the point is the element of geometry, so the element of astronomy is the instant or moment of time ("atomus").[50] In the graduated circles on an astrolabe, the points are made to correspond with numbers in an arithmetic continuum (i.e., the series of all rational and irrational numbers); likewise, elements of a time-series are made to correspond with the points and numbers. It could legitimately be argued that the astrolabe was incapable of such precision,[51] and of course it was possible to debate whether instants have real existence and whether time is discrete or continuous. But a practical astrolabist, such as Chaucer, could simply assert that divisions of circles and hence of time can go on indefinitely:

> But understonde wel that these degres of signes ben everich of hem considered of 60 mynutes and every mynute of 60 secundes, and so furth into smale fraccions infinite, as saith Alkabucius.[52]

Chaucer's citation of Alchabitius here represents an importation, from outside astrolabe literature proper, of matter from an astrologer who, as such, needs limitlessly small units of space and time.[53]

Elections, the other part of astrology significantly related to astrolabes and time, is more fully named by the expression "eleccions of tymes," in conviction that the purpose of taking the ascendant and casting a horoscope is to choose the best times for doing things.[54] The practice was a scandal in at least some courtly circles by the fourteenth century, and by the late sixteenth century it seems to have become a scandal among other classes as well.[55] Underlying the astrological considerations connected with celestial motion is the assumption that the motion controls both the measurement of quantity (time) and the alteration of quality (natural change). Natural things exist and change in time, and both time and change are subject to celestial motions. According to Pseudo-Dionysius' *De divinis nominibus* (6 c., Latin translation ca. 860) these motions define days and nights for us, measure months and years, and determine, number, and hold together the circular motions of times and of the things that are in time.[56] More immediately relevant to late medieval construction of time is a similar formulation from a basic university textbook and clerical manual of that period, John of Sacrobosco's *Compotus*: "Time is an effect of the motion of heavenly bodies, from which issue the measure and numerical value of quantity and the alteration of quality."[57] If one important function of the astrolabe was the measurement of *time*, according to quantity, another and equally important function was the discrimination of *times*, according to quality.

Seasonal times and the months and days of the year correspond to the Sun's annual passage through the signs and degrees of the zodiac. More specialized discrimination of times depends upon establishing the ascendant and on that basis dividing or "equating" the heaven into twelve astrological houses. The earliest known original Western treatise on the astrolabe, written by Raymond of Marseilles in 1141, devotes a chapter to the topic.[58] Henry Bate, another early astrolabist (1246-1310), indicates in his *Magistralis compositio astrolabii* that he has gone so far as to modify the instrument itself in order to achieve greater precision in equating houses for astrological purposes.[59] Messahalla, Chaucer, and Fusoris each present not one but two solutions to the problem.[60] Pèlerin presents only one, but he presents it twice, once in his astrolabe treatise and once in his *Livre de eleccions*, an astrological work which, like his astrolabe treatise, is addressed to Charles V, the owner, incidentally, of a dozen astrolabes.[61] The equation of houses for the election of times is important for the present study because it is a chimerical fabrication but is based on a real point, the ascendant, that fixes it in space and time. It is thus partly conventional and partly natural.[62]

Fixing the mathematical time and the natural point of the ascendant was, as explained near the beginning of this essay, accomplished in a single unproblematic operating of the astrolabe. The next step, dividing the sky for the choosing of times, was less automatic. In 1467 the great mathematician Regiomontanus took on the problem of how it was to be done, and in the course of his work he summarized various methods that were then in common use.[63] His own preferred method was generally adopted in the Renaissance, but in the late Middle Ages there were still several competing methods.[64]

The important point for us is not just that house-division was debated but that, being in part a qualitative human construction, it was in principle debatable. As Ptolemy had earlier explained, the science of stars according to quantity (i.e., astronomy) was characterized by the highest degree of certainty, whereas the science of the Stars according to quality (astrology) was not, because "every science that deals with the quality of its subject-matter is conjectural and not to be absolutely affirmed."[65] A basis for later scholarly understanding of this conjectural use of the astrolabe was laid in the twelfth century by certain remarks made under the heading "De astronomia" in the *De divisione philosophiae* of Dominicus Gundissalinus.[66] It was reinforced when the remarks were repeated, with variations and elaborations, in several "introductions to philosophy" produced in the Parisian arts faculty in the thirteenth century, from which milieu they spread to other centers and succeeding centuries.[67] Gundissalinus writes of a science of the stars that describes their courses and positions after the opinions of men for the purpose of gaining knowledge of times ("sciencia, que cursus et posicionem stellarum secundum hominum opinionem describit ad temporum noticiam"). These "times" are opportunities, segments of time independent of man but carved out and selected by him at will as being favorable to his actions.[68]

Gundissalinus draws a pointed contrast between "astronomia," which is "secundum rei veritatem," and "astrologia," which is "secundum opinionem hominum."[69] In medieval usage, to say that something is according to opinion does not necessarily mean that it has no basis in reality; it often means that it is based in physical reality, which is transmutable and contingent and therefore lacks the certainty and determinacy attaching to mathematics.[70] An extreme version of this view of physical reality leads to the conclusion that "the truth of opinion ... does not depend on some determinate mode of being in reality but rather on what appears to the knower" ("veritas opinionis ... non sequatur modum determinatum essendi in rebus, sed potius id quod apparet cognoscendi").[71] In any case we are dealing with physical change in time, and some of Gundissalinus' successors took the trouble to distinguish between *astrologia*-as-physics and *astrologia*-as-superstition, the latter being what we nowadays generally think of as astrology.[72] Even writers who were very suspicious of astrology in our sense of the word -- Nicole Oresme, for example, and Pierre d'Ailly -- allowed that the stars could be read concerning "qualities of times" ("qualitates temporum").[73] The times in question are local, and accordingly Gundissalinus' characterization of the science was later expanded to take into account regions of earth as well as opinions of men ("astrologia" is "secundum situs terrarum et opiniones hominum").[74]

The usual misgiving about prognostication by the stars was that it might imply a necessity of events to the exclusion of free will. This misgiving could be obviated by saying that whereas the past is necessary, the future is contingent.[75] But the sphere of any celestial body contains its whole trajectory at once, and its representation on an astrolabe is a circle, a figure without beginning or end.[76] An astrolabe thus tacitly attributes to time a uniformity and universality in which calculations for the future are no different from the calculations for the present or the past. The astrolabe's real position among the medieval sciences was just after the demonstrative proofs of mathematical astronomy but before the practical science of acting on judgments in physical astrology.[77] In automatically combining the telling of time with the taking of the ascendant, the astrolabe confounds elements from the sciences on each side of it: the science for which the astrolabe was the instrument was "the science of mobile quantity according to regions of earth and the opinions of men for the purpose of declaring for every difference in time the exact effects in these inferiors" ("scientia quantitatis mobilis secundus situs terrarum et opiniones hominum, ad omnium differentiam temporis certos effectus in hiis inferioribus denuntians").[78] Elections of times are decided not by reference to celestial objects alone but by reference to celestial objects and human designs conjointly.

The times that are distinguished in an election arise from two conjoined but distinct acts: the act of knowing mathematical time by motions of celestial bodies according to truth ("de motibus superioribus corporum secundum veritatem") and the act of subjective perception or conception in men ("opiniones hominum").[79] In this they bear a provocative resemblance to some late-medieval constructions of time itself, particularly that of William of Okcham.[80] Ockham cites the classic Aristotelian definition of time as "the number of motion in the primum mobile" but gives it his own reading by saying that "the word *time* signifies the continuous and uniform motion of the primum mobile and also signifies or consignifies the soul numbering the before and after of this motion and the motion in between."[81] A difficulty with such a view had always been that even a person who did not directly perceive the motion of the primum mobile nevertheless had an apprehension of time. Ockham solved the problem by making use of Averroës' argument, based on Peripatetic astrology, that since all terrestrial change depends on celestial motion, our very experience of living in a transmutable world is an indirect apprehension of that motion and hence of time.[82] Ockham's exposition is expressed in logical terms, so that in effect it is Averroës' solution without the astrology that is Averroës' real support.[83] It is arguable, therefore, that Ockham's construction of time had hidden roots in the astrologers' construction of times. The constructions, in any case, share the status of being mixed objects, "simul naturalia et artificialia": like wine or statues, they are founded in nature but are transformed artificially.[84] In what may be seen as a further development in this line of thinking, Nicholas of

Cusa in his *De ludo globi* (1462-63) treated time with its years, months, and hours as the measuring soul's instrument for marking motion.[85] He saw the astrolabe as also such an instrument, likewise created in the soul though expressed or unfolded in sensible matter.[86]

If time and the astrolabe were instruments of the soul, then time was a combination in the soul of what is known in the intellect and what is perceived by the senses. Intellectually, with reference to a mathematical astronomy of circles, points, and angles, time had a Newtonian equability. Physically, as inflected for qualitative variation, it could be apprehended in individuated segments, of any extent from instants to seasons to ages, shaped to accomodate human perceptions and purposes. In addition to marking natural days and years and the conventional parts of each, the astrolabe could be used to carve out ad hoc experiential times analogous to the ones we have in mind when we speak of a lucky day, hard times, the Victorian period, the Sixties. The astrolabe, in short, contained both the time that waits for no man and the times that try (and are shaped by) men's souls.

In the edition of Johannes Kepler's *Tabulae Rudolphinae*, published at Ulm in 1624, there is a frontispiece designed by Kepler himself that shows the "House of Astronomy" with a dome supported by pillars representing the great astronomers. On the pillar representing Ptolemy hangs an astrolabe, in accordance with medieval astrolabists who attributed the origin of their instrument to "the venerable Ptolemy."[87] On the pillar representing Tycho Brache (1546-1601), the last great naked-eye observer, hangs a rather large quadrant, incorporating a radius about three times that of the astrolabe on Ptolemy's pillar. This difference in size is appropriate because such a quadrant was itself a development from the astrolabe in an attempt to achieve greater accuracy in observation by increasing the size of that part of the instrument by which observations are made.[88] In improving observational accuracy, however, the quadrant sacrificed the astrolabe's representational function[89] and most of its computing capabilities. Kepler's frontispiece thus presents a little symbolic history of the astrolabe. The instrument had a long and brilliantly successful career, going back in one form or another as far as Ptolemy, but in the new observational astronomy of Tycho and Kepler it had little role to play. In its perfected state in the late Middle Ages it was indeed an observational instrument, with enough accuracy to be the best device there was for telling time. But to a far greater extent it was an instrument for thinking about celestial spheres and motion and time. In relating qualitative times to impartial time, it helped people to distinguish the two, to abstract time from particular living rhythms and changes, and to see a year or a day or hour as a unit of time and not just as an episode of experience.

Southwest Texas State University

Notes

1. A readily available description of the astrolabe, with pictures and diagrams, is J.D. North's "The Astrolabe," *Scientific American* 230 (January, 1974), 96-106. On the development of the instrument, see Otto Neugebauer, "The Early History of the Astrolabe," *Isis* 40 (1949), 240-56. An excellent general account, with pictures and diagrams, is *The Time Museum*, vol. 1 (*Time Measuring Instruments*), part 1 (Astrolabes), by A.J. Turner (Rockford, IL: Time Museum, 1985).

2. Augustine canvasses the questions comprehensively in *Confessiones* XI. See James McEvoy, "St. Augustine's Account of Time and Wittgenstein's Criticisms," *Review of Metaphysics* 38 (1984), 547-76, esp. 566, where Wittgestein is quoted as saying that "someone who is engaged in measuring time will not be bothered with the nature of this problem (what is time?)."

3. The principle is often repeated. I quote from John of Saxony's canons (ca. 1327) to the Alfonsine Tables, ed. Emmanuel Poulle, *Les Tables Alfonsines* (Paris: CNRS, 1984), 30: "Tempus est mensura motus primi mobilis, ut vult Aristoteles quarto *Physicorum*" (*Physics* IV, 12 220b).

4. Ed. by R. T. Gunther in *Chaucer and Messahalla on the Astrolabe*, vol. 5 of *Early Science in Oxford* (Oxford: Oxford UP, 1929).

5. Ed. by Edgar Laird and Robert Fischer in *Pèlerin de Prusse on the Astrolabe* (Binghampton, NY: Medieval and Rensaissance Texts and Studies, 1995).

6. Ed. by John Reidy for *The Riverside Chaucer*, ed. Larry D. Benson (Boston, MA: Houghton Mifflin, 1987).

7. Ed. by Emmanuel Poulle in *Un Constructeur d'instruments astronomiques au XVe siècle, Jean Fusoris* (Paris: Champion, 1963).

8. Messahalla, op.cit., II.3.

9. *Practique*, II.3.

10. Chaucer, op.cit., II.3.

11. Fusoris, op.cit., II.2. As explained by Theodorus Meliteniotes (14 c.), *Tribiblos astronomique*, ed. and trans. Regine Leurquin (Amsterdam: Gieben, 1990), 223, when we have set the rete, we have *automatically* located the four "centers" (*kentra*) on which the principal houses are based. (The centers are the ascendant and its opposite and the "culmination" or mid-sky and its opposite.) This is so, he explains on page 221, because once the instrument is set it will have the same arrangement as the universe, for the instrument is constructed in representation of the universe.

12. *Sententie astrolabii*, ed. J.M. Millás Vallicrosa in *Assaig d'historia de les idees fisiques i mathematiques a la Catalunya medieval*, vol. 1 (Barcelona: Institució Patxot, 1931), 275.

13. J.M. Millás Vallicrosa, "Translations of Oriental Scientific Works," in Guy S. Métraux and François Crouzet, eds., *The Evolution of Science* (New York: New American Library, 1963), 138-46; Arno Borst, *Astrolab und Klosterreform an der Jahrtausendwende* (Heidelberg: Winter, 1989), 43ff; also Charles Homer Haskins, *Studies in the History of Medieval Science* (Cambridge, MA: Harvard UP, 2nd ed., 1927), 9-10, 114-5; Robert W. Southern, *Medieval Humanism* (New York: Harper & Row, 1970), 166-7.

14. Paul Kunitzsch, "The Treatise on the Composition and Use of the Astrolabe Ascribed to Messahalla," *Archives internationales d'histoire des sciences* 31 (1981), 54, and note 55.

15. Willy Hartner, *Oriens-Occidens* (Hildesheim: Olms, 1968), 297; Harold N. Saunders, *All the Astrolabes* (Oxford: Senecio, 1984), 20.

16. Borst, *Astrolab*, 47.

17. Emmanuel Poulle, "L'Astrolabe médiéval," *Bibliothèque de l'école des Chartes* 112 (1954), 89.

18. Jaques Le Goff, *Time, Work, and Culture in the Middle Ages*, trans. Arthur Goldhammer (Chicago: U of Chicago P, 1980), 44.

19. Astrolabes were typically furnished with several different coordinate plates that could be inserted as approproiate to different geographical latitudes. The last chapter of Fusoris's treatise describes his attempt to construct one plate that would serve for all latitudes "pour eviter la multitude de tables (i.e., coordinate plates) qu'il fauldroit pour servir tous les climas et leurs parties." Such "universal" astrolabes became common in the Renaissance, and one had been described in the late 13c. by Petrus Peregrinus de Maricourt, *Nova compositio astrolabii particularis*, in Petrus's *Opera*, ed. Lois Sturlese and Ron B. Thompson (Pisa: Scuole Normale Superiore, 1995).

20. Saunders, *Astrolabes*, 20.

21. *Traité de l'espere*, Oxford, St. John's College, MS 164, fol.6v: "n'est pas fait ne causée de nature, mais de volenté selonc raison d'aritmetique."

22. Messahalla, op.cit., II8.

23. *Practique*, II.6.

24. Chaucer, op.cit., I.16, and II.11; Fusoris, op.cit., I.1.

25. Poulle, "L'Astrolabe,", 94.

26. As already indicated, the "Messahalla" treatise is an editors' product to begin with. The incomplete state in which Chaucer left his treatise tempted some copyists to add to it.

27. *Practique*, Prologue: "chascune personne pouvre pluseurs autres proffis de lui meisme orderner et entendre."

28. Chaucer, op.cit., II.14, and see J.D. North, *Chaucer's Universe* (Oxford: Clarendon P, 1988), 67-8. Fusoris, op.cit., II, prologue, suggests that the astrolabe is good for users "par maniere d'aucune recreacion et eulx esbatre."

29. *Physics* IV, 14, 223a.

30. Michael Scot (early 13 c.) makes the application in his commentary on the *Sphere* of Sacrobosco, ed. Lynn Thorndike in *The "Sphere" of Sacrobosco and Its Commentators* (Chicago: U of Chicago P, 1949), 248-9: "omne demonstrationes que fiunt in spera materiali similiter debent intelligi in mundo sensibili, quia abstrahentium non est mendacium." Scot may be referring to an armilliary sphere, but the argument applies to an astrolabe as well, for an astrolabe has the form of "a sphere spread out flat" (Messahalla, op.cit., I, prohemium: "spera que fuerit in extensa plano"). Fusoris, op.cit., I.2, says that one may imagine an astrolabe as the shadow cast on a wall in a certain way by an armilliary sphere.

31. Poulle, *Constructeur*, 91. Thomas Bradwardine (14 c.), *Geometria speculativa*, ed. George Molland (Stuttgart: Steiner, 1989), 20, says that the astrolabe serves the practice of geometry, that geometry concerns relative properties like equality and inequality, and that whether such properties have seperate existence is not a concern of geometry.

32. Michael Scot, in his commentary on Sacrobosco, in Thorndike, op.cit, 323: "Tempus mathematicorum, id est, tempus astronomorum."

33. This astronomer's truism comes from Ptolemy, *Almagest*, I.1, whence (with reinforcement from Aristotle, e.g. *Posterior Analytics*, I, 12, 77b) it is repeated by medieval writers, including Robert Grosseteste, *Commentarius in Posteriorum analyticorum Libras*, I.19, and Thomas Aquinas, *Expositio super librum Boethii De Trinitate*, quaest. VI, art. 1, both of whom cite Ptolemy.

34. Emmanuel Poulle, "Instrumentation astronomique médiévale," in Bernhard Ribémont, ed., *Observer, lire, écrire le ciel au moyen âge* (Paris: Klincksieck, 1989), 255.

35. Le Goff, op.cit., 48.

36. *Sententie astrolabii*, in Millás Vallicrosa, *Assaig*, p. 275; *Capitula horologii Regis Ptolomei*, ibid., 280; *De utilitatibus astrolabii*, ed. R.T. Gunther, *Astrolabes*, 409; Messahalla, op.cit., II.3; etc.

37. Dominicus Gundissalinus, *De divisione philosophiae*, ed. Ludwig Baur (Munich: Beiträge zur Geschichte der Philosophie des Mittelalters, 4: 1903), 120: "astrolabium est quo comprehendimus certas horas diei et noctis"; repeated in *Philosophica disciplina* (ca. 1245), ed. Claude Lafleur, *Quatre introductions à la philosophie au xiii siècle* (Montréal: Institut d'Etudes Médiévales; Paris: Vrin, 1988), 272. On the effects of astronomical hours on urban time-keeping, see Arno Borst, *The Ordering of Time*, trans. Andrew Winnard (Chicago: U of Chicago P, 1993), 92-7. The dials of many early clocks were, in effect, machine-driven astrolabes. In certain expensive shops, one can still buy astrolabe-faced clocks.

38. Haskins, *Medieval Science*, 28-9.

39. His prologue speaks not only to "Lyte Lowis" but also to "every discret persone that redith or herith this litel tretis."

40. For example, the Bodleian's MS Rawlinson D.3 is handsome enough to suggest aristocratic patronage; its Ashmole 393 is a penurious abridgment, closely written on paper; its Bodley 619, with its accurate technical marginalia in Latin, suggests a learned readership. All three date from the fifteenth century.

41. Le Goff, op.cit., 34-5.

42. Stuart Jenks, "Astrometeorology in the Middle Ages," *Isis* 74 (1983), 185-202.

43. Ptolemy, Almagest, I.1; Christine de Pizan *Le Livre des fais et bonnes meurs du sage roy Charles V*, vol. 2, ed. S. Solente (Paris: Champion, 1940), 17, quoting Ptolemy.

44. Fusoris, op.cit., prologue.

45. Nicole Oresme's two treatises against judicial astrology, both ed. by G.W. Coopland in *Nicole Oresme and the Astrologers: A Study of His "Livre de divinacions"* (Cambridge, MA: Harvard UP, 1952), are directed to "principes et magnates" and to "personnes d'estat comme sont princes et seigneurs auxquels appartient le gouvernment publique."

46. *Practique*, II.3; cf. Messahalla, op.cit., II.3; Chaucer, op.cit, II.3; and Fusoris, op.cit., II.2.

47. *Practique*, II.1.

48. To give but one example, Jean Froissart, who was present in Bordeaux when Richard II was born there, gives the time of the birth as "about 10 o'clock," but a contemporary horoscope discussed by J.D. North, *Horoscopes and History* (London: Warburg Institute, 1986), 141, shows that the horoscope-maker used astrological computations to "rectify" the time of Richard's birth and arrived at a figure differing from Froissarts's by three hours.

49. See Augustine, *De civitate dei*, V.3. Robert W. Southern, *Robert Grosseteste: The Growth of an English Mind in Medieval Europe* (Oxford: Clarendon, 1986), 103, argues that it was the arrival of the Astrolabe in the West, with its theoretical precision, that overcame Augustinian arguments concerning the destinies of twins.

50. Prologue to a tenth-century treatise, probably by Llobet of Barcelona, ed. Millás Vallicrosa, *Assaig*, 272. The same assertion is made in the twelfth century by Hugh of St. Victor, *Didascalion*, VI.14.

51. The argument was made in 1396 by Coluccio Salutati, *De fato et fortuna*, quoted by Charles Trinkhaus, "Coluccio Salutati's Critique of Astrology in the Context of His Natural Philosophy," *Speculum* 64 (1989), 65. Late medieval astrolabes were accurate to within 2-5 minutes of time, according to A.C. Crombie, "Quantification in Medieval Physics," in Sylvia Thrupp, ed., *Change in Medieval Society* (New York: Appleton-Century-Crofts, 1964), 203.

52. *Astrolabe*, I.8. Chaucer goes on to relate minutes of an arc to minutes of time: sixty minutes of the arc of the celestial equator equals four minutes of time. Twenty minutes of time, moreover, equals a "mile wey," i.e., the time required to walk a mile.

53. The reference is to a passage near the beginning of *differentia prima* of Alchabitius's *Introductorius ad magisterium judiciorum astrorum*, of which there is no recent edition. It is a tenth-century Arabic astrological work translated into Latin in 1144 and into French and English in the late fourteenth century.

54. Chaucer, *Astrolabe*, II.4, and cf. *Philosophica disciplina* in Lafleur, op.cit., 286: "Oroscopium est consideratio temporum in rebus agendis."

55. Philipe de Mézières (writing in 1387), *Le Songe du Vieil Pelerin*, vol. 1, ed. G.W. Coopland (Cambridge: Cambridge UP, 1961), 598, complains that lords are unwilling even to don a new robe or leave the palace without first making an election. Joseph Hall (writing ca. 1597), *Virgidemiarum*, II.7, writes, "Who pares his nails or libs his swine/But he must first take counsel of the sign?"

56. *De divinis nominibus*, IV.12, in Celsus Pera, ed., *St. Thomas Aquinas In librum beati Dionysii de divinis nominibus* (Turin: Marietti, 1950), 96.

57. "Tempus igitur est effectus corporum supercelestium motus, ex quo quantitatis dimensionem et numerum et qualitates alterationem sortitur," quoted by Olaf Pederson, "In Quest of Sacrobosco," *Journal for the History of Astronomy* 6 (1985), 220. Pederson notes (280) that the text seems corrupt here, but the sense seems available. On Sacrobosco's tomb in Paris, incidentally, there was a carving of an astrolabe.

58. Raymond's treatise, ed. Emmanuel Poulle, "Le Traité d'astrolabe de Raymond de Marseille," *Studi medievali*, 3rd series, 5 (1964), 866-909, is not directly dependent on an Arabic source. Chapter 18 is "De domibus per astrolabium diagnoscendis."

59. Gunther, ed. *Astrolabes*, vol. 2, 369. The work was printed by Erhardt Ratdolt in 1485.

60. Messahalla, op.cit., II.37, 38; Chaucer, op.cit., II.36, 37; Fusoris, op.cit., II.2,6.

61. *Practique*, II.16. The relevant section of the *Livret* is printed as Appendix C in Larid and Fischer, eds., *Pèlerin*. Charles's astrolabes are catalogued in Jules Labarte, ed., *Inventaire du mobilier de Charles V, roi de France* (Paris, 1879).

62. In the fourteenth century, Nicole Oresme, *Traité de l'espere*, chapter 9, asserted that only natural points ("poins ... fondée sus la nature") are meaningful. In the sixteenth century, Jean-Baptiste Morin, *Astrologica gallica*, as reported by Jim Tester, *A History of Western Astrology* (New York: Ballantine, 1987), 237, insisted that house-divisions are not entirely man-made, that at least some, including the ascendant, are based on natural points at determinable times.

63. *Tabulae directionum*, sec.14. The work was not printed until 1490 (by Erhardt Radtolt, Augsburg). It is repr. in Felix Schmeidler, ed., *Regiomontari opera collectanea* (Osnabrück: Zeller, 1972).

64. A modern account of the old methods is given in J.C. Eade, *The Forgotten Sky* (Oxford: Clarendon P, 1984), 42-7.

65. *Tetrabiblos*, op.cit., I.2, on astrology; cf. *Almagest*, op.cit., I.1, on astronomy.

66. Ed. Baur, op.cit., 119-21.

67. Lafleur, op.cit., 158-61.

68. Max Lejbowicz, "Le Choc de traductions Arabo-latines du xii siècle et ses consequénces dans la spécialisation sémantique d'*astrologia* et d'*astronomia*," in Martine Groult, ed., *Transfert de vocabulaire dans les sciences* (Paris: CNRS, 1988), 225.

69. *De divisione philosophiae*, in Baur, op.cit., 121.

70. See, for example, Robert Grosseteste, *Commentarius in Posteriorum analyticorum libros*, ed. Pietro Rossi (Florence: Olschki, 1981), I.19, ll. 84-9, and I.17, ll. 350-2; see also Stephen P. Marrone, *William of Auvergne and Robert Grosseteste* (Princeton: Princeton UP, 1983), 222-3.

71. This is the "Heraclitizing" view reported by Thomas Aquinas, *In duodecim libros Metaphysicorum expositio* IV, lc.12, 683, trans. John P. Rowan, *Commentary on the Metaphysics of Aristotle*, vol. 1 (Chicago: Regnery, 1961), 273.

72. *Philosophica disciplina*, in Lafleur, op.cit., 273. Cf. Robert Kilwardby, *De ortu scientarum* (ca. 1250), ed. Albert P. Judy (Toronto: Pontifical Institute of Mediæval Studies, 1976), 32-3, where star-study is similarly divided into "vere mathematica" (quantity), "naturalis" (quality), and "quasi naturalis" (lies and superstition).

73. Oresme, *Contra judiciarios astronomos* (ca. 1360), and D'Ailly, *De falsis prophetis* (after 1409), both ed. by Coopland, *Oresme*; see 133 and 146. Both use the phrase "qualitates temporum." Oresme cites Augustine on what is allowable "ad judicia futurorum." D'Ailly, *Vigintiloquium*, verb. 3, ed. Erhardt Radtolt (Venice, 1490), cites both Augustine and Ptolemy on what is allowable and says "concordat igitur theologia & astronomia."

74. *Accessus philosophorum*, ed. Lafleur, op.cit., 219.

75. Henry of Langenstein (d. 1397), *Expositio prologi Bibliae*, quoted in Steneck, "Arbor Scientarum,", 259, lists common questions "ut de praedestinatione, de contingentia futurorum, de necessitate praeteritorum...." In his *Tractatus contra astrologos*, ed. Herbert Pruckner, *Studien zu den astrologischen Schriften des Heinrich von Langenstein* (Leipzig: Teubner, 1933), 160, Henry says it is easy to trace events backward to celestial causes, but not to foretell future events from present celestial conditions.

76. Averroës, *Commentarium medium in Aristotelis De generatione et corruptione libros* I.12, ed. F.H. Fobes (Cambridge, MA: Medieval Academy of America, 1956), 154: "tempus igitur est numerus motus continui; sed continuum est circulare; ergo tempus est numerus circularis motus." In the pervasive schema of sciences transmitted to the Middle Ages through Boethius' *De Trinitate*, astronomy can be viewed mathematically, as concerning objects considered without matter (e.g., circles) or physically, as concerning objects with with matter and motion. The astrolabe does both, marking tine mathematically and times physically.

77. See Campanus of Novara, *Theorica planetarum* II.40-1, ed. Francis S. Benjamin and G.J. Toomer, *Campanus of Novara and Medieval Planetary Theory* (Madison, WI: U of Wisconsin P, 1971): also Laird and Fischer, Pèlerin, 10-12.

78. *Compendium circa quadrivium*, ed. Lafleur, op.cit., 363.

79. *Accessus philosophorum*, ed. Lafleur, op.cit., 219.

80. I refer to *Tractatus de successivis*, ed. Philotheus Boehner (St. Bonaventure, NY: Franciscan Institute, 1944), which is compiled, by Ockham or someone else, from Ockham's commentary on Aristotle's *Physics*. Part 3 of *De successivis* constitutes a handy "Tractatus de tempore."

81. *De successivis*, 110: "hoc nomen tempus significat primum motum continuum et uniformem, et etiam significat vel consignificat ipsam animam numerantem prius et posterius illius motus et motum medium."

82. *De successivis*: Averroës, says Ockham, "dicit, quod non sentimus nos esse in esse transmutabili nisi ex transmutatione caeli" (107); "intendit commentator, quod quicumque sentit quecumque motum, potest sentire tempus.... Aliquis non videns corpus caeleste moveri, apprehendit aliquem motum exteriorum vel imaginatur motum" (105).

83. Pierre Duhem, *Medieval Cosmology*, ed. and trans. Roger Ariew (Chicago: U of Chicago P, 1985), 315.

84. Gundissalinus, *De divisio*, 10-11; and see Lejbowicz, "Le Choc des traductions," 266.

85. *De ludo globi*, ed. Pauline Moffit Watts (New York: Abaris, 1986), 106): "Sic annus, mensis, horae sunt instrumenta mensurae temporis per hominem creatae. Sic tempus cum sit mensura motus mensurantis animae est instrumentum."

86. Ibid.: "Creat anima sua inventione nova instrumenta ut discernat et noscat: ut Ptholemeus astrolabium.... Neque ex aliquo extrinseco inventores crearunt illa: sed ex propria mente. Et explicarunt enim in sensibili materia conceptum."

87. Pèlerin, *Practique*, prologue: "le tresreverent Ptholomee."

88. Emmanuel Poulle, "Le Quadrant nouveau médiéval," *Journal des savants* (April-June, 1964), 149.

89. John Blagrave's attempt to preserve the astrolabe's representational function in a post-Copernican world is described in *his Astrolabium Uranicum Generale* (London, 1596, facsimile repr. by Da Capo Press, 1972). His instrument represents the celestial sphere as fixed, with moveable coordinates turning over it.

A Learned King Enthralls Himself: Escapement and the Clock Mechanisms in Alfonso X's *Libro del saber de astrologia*

Anthony J. Cárdenas

For Robert I. Burns, S.J.

What could motivate anyone to sponsor learning five different categories -- history, law, religion, science, and even recreation?[1] The reasons for patronage are undoubtedly as numerous as the sponsors and the individual works sponsored themselves. In the case of Alfonso X, whose sobriquet "el Sabio" or "the Learned" is well deserved, glimpses of his motives can be found within his own writings.[2] For one paramount motive, one might, for example, consult the following passage from the *General estoria* II:

> Et dizen otrosi que cada vno quanto mas a del saber e mas se llega a el por estudio, tanto mas aprende, e creçe, e se llega por ende mas a Dios, como monte que va suso al çielo.[3]
>
> [And they also say that the more one has of knowledge and the more one approaches at it through study, the more one learns and grows and the more one nears God because of it, as a mountain that goes upward toward heaven.]

Another reason lies in a tendency noted by Francisco Rico in which the Learned Monarch views the great characters of his Universal History, i.e., in his *General estoria* as prefigurations of himself.[4] Similarly, Charles F. Fraker has registered an effort to bolster his regal position and his imperial endeavors by means of his *Estoria de España.*[5]

Other rationales for his sponsorship appear in his various works. The very nature of the centralizing tendencies of his *lex legium*, his *Siete partidas*, it might be conjectured, smacks of preparation for his quest for Holy Roman Emperorship. In his *Cantigas de Santa Maria*, on the other hand, we learn in the prologue that he wishes to be the Virgin's own troubadour, who forsakes all other women to serve only the Mother of God:

> quero seer oy mais seu trobador,
> e rogo-lle que me queira por seu

Trobador e que queira meu trobar
reçeber...
... e ar
querrei-me leixar de trobar des
por outra dona, e cuid' a cobrar
per esta quant' enas outras perdi.[6]
[I want today to be her Troubadour
and I beg her that she will want me for her
Troubadour and will want my song
to receive...
... and also
that I will want to stop singing immediately
for another woman, and I hope to recover
through this one all that I lost on the others.]

Entering the good graces of the Mother of God is assuredly to secure a "piece of the rock," as it were, a way to assure success both in this and in the afterlife. In the treatise dated 1283, one year before his death, the *Libro de açedrex & dados & tablas* (*Book of Chess and Dice and Board Games*), we read introduction that such games are useful for, among others, those who cannot partake in jousting, namely, women, children, those incarcerated, and the *elderly*:

> estos iuegos ... se fazen tanbien de noche como de dia, e por que las mugieres que non caualgan e estan encerradas an a usar desto, e *otrossi los omnes que son uieios e flacos*, o los que han sabor de auer sus plazeres apartadamientre ... o los que son en poder ageno assi como en prision o en catiuerio o que uan sobre mar, e comunalmientre todos aquellos que han fuerte tiempo, por que non pueden caualgar nin yr a caça ni a otra parte, e han por fuerça de fincar en las casas e buscar algunas maneras de iuegos con que hayan plazer e se conorten e no esten baldios.[7]
>
> [these games ... are played at night as well as in the daytime, and because women who do not mount horses and are enclosed ought to use these, and also men who are old and weak, or those who feel like taking their pleasures removed ... or those who are in someone else's power, as if in prison or in captivity, or those who travel on the sea, and naturally all those who endure inclement weather, because they cannot mount horses, nor go hunting nor go to any other place, and they must remain in their homes and must look for kinds of games with which they may have pleasure and console themselves and not be idle.]

In short, it might be said that Alfonso's motives for sponsoring learning were practical. Through scholarship, in one form or another, he was attempting to protect, bolster, advance his position as King and as potential Holy Roman Emperor.

Now, recalling the notion quoted above that the learned man approximates divinity, and looking at his most prolific area of sponsorship, science (read "astrological") treatises, it cannot surprise that the legend of the bumbling astrologer king,[8] who in a moment of hubris is purported to have declared: "Had I been present at the Creation, I would have given some useful hints for the better ordering of the universe..."[9] should shroud historical fact regarding the Learned King.[10] In his science treatises, motives are plentiful, but among these various treatises, one finds notions such as the need to know the appropriate time to wage war, to be known as a sage; to be a king feared by others and obeyed by all, to be well-loved as a king, and even how to make bedbugs and fleas flee.[11] Alfonso's interest in time and its measurement, then, was markedly different from the monastic interest, not so much in manner -- measurement by observation of stellar movement[12] -- as in motive. Monasteries needed to determine canonical hours, and Alfonso needed to determine the most opportune time for myriad political purposes. It will not suprise, then, that the devices of measurement in his *Libro del saber de astrologia* served to keep track of the astrological configurations in the heavens.

What this study shall examine briefly concerns that portion of Alfonso X's own *Libro del saber de astrologia*, formerly and still incorrectly cited as *Libros del saber de astronomia*,[13] which pertains to horologes and specifically the matter of escapement. Along the way, Alfonso's enthusiasm emanating from the ingenuity of his devices will become patent.

By way of introduction, Alfonso X's *Libro del saber de astrologia* is primarily a scientific anthology comprising sixteen treatises, a thirteenth-century "Everything-You-Wanted-To-Know-About-The-Construction-And-Use-of-Medieval-Instruments-of-Astronomical-Measurements. It was begun in its final form in 1276, was in progress in 1277, and was finished at some indeterminate time after that.[14] The text itself survives in its original form, that is, in the codex emanating from Alfonso's Royal Scriptorium, today MS 156 of the Central Library of the University of Madrid. The content of the anthology falls logically into four units: The first unit comprises a star catalogue indebted to Ptolemy's *Almagest*; the second group includes nine treatises concerning the construction and use of instruments employed in direct astronomical observation -- astrolabes of one sort or another, and instruments derived therefrom; the third unit, the focus of this essay, contains five treatises on various kinds of clocks; and the last section, like the second, consists of a treatise, an instrument of direct observation called "atacyr," an astrolabe-like instrument calibrated for and used specifically in astrological reckonings.[15]

The clock unit, folios 172r to 197v of MS 156, although incomplete at the beginning and at the end,[16] deals with five clocks: 1) the standard Graeco-Roman sundial, 2) a clepsydra, 3) a mercury clock, 4) a candle clock, and 5) a construction called "palace of the hours" which might be considered a reverse sundial since the principles involved are similar, with the exception that, unlike the sundial, in which the shadow of a gnomon is used to mark solar passage, in the "palace of the hours," a ray of light entering one and only one of the twelve appropriately placed windows does the same duty as the shadow of the gnomon. The "palace of the hours" finds its New World analogue in the oft-acclaimed Mayan sun temples.

Although any number of potential facets lend themselves to scrutiny, as stated, this study is limited to the question of escapement and to Alfonso's reaction to these works.[17] Escapement in this study does not mean that of "locking and impulse" as defined by J. Eric Haswell[18] nor does it refer to the "two classes (of escapement), namely, 'recoil' and 'dead-beat'" as he continues to clarify.[19] Instead, escapement is that as defined by Lynn White, Jr.: "a means of obtaining an even flow of energy through the mechanism."[20] Thus neither sundial nor palace have to negotiate by a mechanism an even flow of energy; they are consequently excluded from further discussion.

This study will suggest that the ingenuity manifested in the various solutions to the problem of harnessing gravity places Alfonso in the mainstream of the Graeco-Arabic technology. Finally, the elaborateness of the instruments described within and the superlative and skilled craftsmanship required for their production hint that these clocks, although functional or at least probably functional, were likely intended to be ingenious showpieces for the nobility of his court (for who else would have access to them?), and certainly to be testimony to his great learning.

This study focuses, then, on the interior three clocks of this unit -- the clepsydra, the mercury clock, and the candle clock -- which, being more complex, share one common element: gravity as their power source. For the clepsydra it is gravitational pull on the water itself, and for the other two horologes, on weights which form part of the device.

Because these three mechanisms are solar-independent, they are always functional -- especially when they might be needed the most -- when the heavens were hidden from direct observation. The designs of all three -- and especially that of the mercury clock and that of the candle clock -- provide ingenious solutions to the medieval technological problem of escapement. As Lynn White notes:

> The most notable study of differential speeds, however, was carried on in relation to that ingenious medieval invention, the mechanical clock. As has been mentioned, the engineers of the thirteenth century became fascinated by the problem of devising a chron-ometer operated by the force of gravity. The difficulty was to discover an escapement, that is, a means of obtaining an even flow of energy through the mechanism. The previous history of technology offered no precedent for such an effort, save by a flow of water, which was unsatisfactory in a climate where icing was frequent.[21]

Establishing adequate escapement in the clepsydra appears to be precisely Alfonso's primary concern, as well as his motivation for writing the following at the beginning of the treatise on the clepsydra:

> Et por que esta cosa es muy sotil de saber ell ombre todo esto a menos de ueer sol o alguna estrella; por end nos Rey don Aflonso el sobredicho ouiemos sabor delo poner en este nuestro libro. Et lo que fal-lamos escripto en los libros que fizieron los sabios antigos era muy menguado. Et esto es por que ellos foradauan la tinaia o a de seer ell agua en so fondon. & corria ell agua dende en la primera hora; mas que en la segunda. & en la segunda; mas que en la tercera. & segund esta obra salien las horas yguales. desyguales. & non sabian por aquellos sus relogios ell ascendente. ni otras cosas que se pueden entender por ello.... Et nos touiemos por bien de fazer este relogio de otra manera de guisa que non aya y yerro niguno....[22]
>
> [And because the subject is very subtle for man to know unless by observing the sun or some star, for that reason, we the aforesaid King don Alfonso desire to put this in our book. And what we found written in the books by the ancient sages was very deficient. This is owing to their drilling the vessel, from which the water was to exit, near the bottom. Thus more water exited in the first hour than in the second, and more in the second than in the third. And so the hours which were to be equal were not; and they could not know by those horologes the ascendant nor other things that can be understood by such .... And we thought it well to make this clock another way so that there would be no error at all....]

Although none of the studies on Alfonso's water clock[23] or on water clocks in general[24] which I have been able to consult discuss the matter of controlling water pressure to provide a constant rate of flow as a problem, surely this problem must have been solved prior to Alfonso's treatise as Lynn White's remarks above suggest.

Even though Alfonso does not address the complication presented by freezing water, as mentioned by Lynn White above, the length of this particular treatise, 11.5 folios, nearly twice the size of the next largest clock treatise, is in large part a result of the detailed discussion necessitated by the delicacy of the main principle on which the functioning of this timepiece is based: its escapement, i.e., the constant and continuous flow of water. To safeguard this flow Alfonso 1) recommends the use of lids to keep dust and other foreign particles out of the water, 2) insists upon the cleanliness of the water and the vessels in which it is contained, 3) suggests using a linen cloth of appropriate weave as a particle-screen, that is, filtered, over various water exits, and 4) encourages the use of ground alum to rarify the water.[25] In common with other water clocks Alfonso's too has an hour marker, i.e., a float on which there rests a panel divided into three hundred and sixty degrees with every fifteen equalling one hour. The float itself is placed inside the receptacle into which the water flows ideally at an even rate. One can view on the panel which rests on the float, the twelve signs of the zodiac quite artistically depicted. The plan, as in the case of the candle clock, as we shall see below, is to have the appropriate zodiacal sign become visible at the moment the actual constellation rises in reality over the horizon. Provisions are also made for knowing apparent time and for calibrating the Roman months.

The other two clocks present unique solutions to the matter of escapement. Both use gravitational pull as their power source but this time on lead weights. Alfonso's mercury clock more than his others has attracted the attention of scholars interested in medieval horologes. H. Allen Lloyd, for example, provides a quite adequate assessment of the mechanism in his book *Some Outstanding Clocks Over Seven Hundred Years 1250-1950*. He writes:

> ... 'stub' pinion of six leaves is affixed to the same arbor as the two drums, and in front; thus, the pinion would also turn once in four hours. Above it, on its own arbor, is the dial of an astrolabe, framed in an oak circle into which are driven thirty-six wooden teeth made of an especially hard kind of oak, i.e., the teeth are at 10 1/2 intervals. As the drum and the 6-tooth stub pinion revolve once in four hours, in six turns, i.e., twenty-four hours, the pinion will turn the

'rete', or perforated front dial of the astrolabe, one revolution. Just as easily could a twenty-four-hour dial have been substituted to register the passing of mean time. Thus, we have here a contemporary illustration of what may be considered the forerunner of the mechanical clock as we know it today. As mentioned above Plate I shows the Mercury clock illustrated; on the right is seen a large drum round which is coiled a rope having a weight hanging down at the end. On the same arbor (or shaft) of the drum is fixed another smaller one, divided into twelve compartments, with small holes in the dividing walls. The lower half is filled with mercury. As the weight tends to pull the large drum round, it raises the mercury until it counter-balances the driving weight, which can then only fall very slowly as the mercury flows through the small holes in the dividing walls. Thus the uniform motion of the drums is dependent on the viscosity of the mercury in the inner drum. Regulation was effected either be altering the weight, or the diameter of the drum, or both. The period of revolution was to be four hours.

The bar seen in the drawing indicates that the lower, it has come to us in a thirteenth-century MS., but it may well have been copied from one, centuries earlier; we shall probably never know.[26]

There is nothing in the text by Alfonso to either affirm or contradict Lloyd's closing statement regarding the originality of this treatise. Alfonso does, however, state the following:

Poren nos Rey don Alfonsso el sobredicho mandamos al dicho Rabiçag que fiziesse un libro de cuemo puede fazer este relogio por la arte del libro que fizo yran el philosopho en que fabla de cuemo se pueden alçar las cosas pesadas.[27]

[And for that reason, we, King Alfonso the aforesaid, ordered the said Rabiçag to make a book on how one can make this clock by the art of the book that Yran the philosopher made in which he speaks of how one can raise heavy things.]

If "Yran el philosopho" is in fact Hero of Alexandria as Alfred Wegener suggests by a query,[28] consulting whatever survives of his work, probably his *Mechanics*, may resolve the matter of originality. Until such information can be examined, it is foolhardy to doubt any originality or to claim total originality for this mechanism. In his discussion above of the mercury clock, Lloyd further notes the possibility of attaching a modern dial and thereby measuring mean time, i.e., substituting the astrolabe with a modern, 24-hour, clock's face as we know of it today.[29] In a sense, Alfonso provided for this, although it is not depicted in his illustration. He suggests the proper arrangement of four bells so that, as the shaft turned, another appropriately calibrated cog wheel would strike the bells at regular intervals. However, that this addition of bells is not depicted in the final diagram of this mechanism may suggest the secondary importance of this device as a timepiece. Its primary function, as an anaphoric clock, was to reflect the motion of the heavens.[30]

Interest in this mercury clock has been such that only a few years ago A. A. Mills described a project to recreate it only to discover that said clock was not very reliable, and to conclude that it is quite probable that neither the author of the Arab original nor the translator had ever seen the clock they attempted to describe, and that Alfonso's work had not impact on the evolution of the clock in the Western world.[31]

It is, nevertheless, interesting to note at this point, that Alfonso uses an Arabic-derived Spanish term, "aceñia," to denote a cog wheel in this contrivance. "Aceñia" then and today means "water wheel." The perennial problem of technologically adequate terminology is hereby resolved by adapting one term for another use.[32] At any rate, Lynn White again provides evidence for use of a cog wheel in another clock which antedates Alfonso's by some twenty-five years. He is quick to note that employment of cogs was no paltry achievement. Although he must interpret the mechanism of this 1250 clock from a schematic miniature, he states that it is the "first instance of the weight drive in a machine other than the trebuchet, the second example being that of the closely related clock of Alfonso X a quarter-century later."[33]

Now, the final clock of our discussion, is the candle clock. The burning of graduated candles is reported to have existed ever since there were candles. One source notes this kind of clock in the eight century. King Arthur of England reportedly used a four-hour candle to divide his working day. In a subsequent development of King Arthur's graduated candle a tiny bell was embedded in the candle so that at about the chosen hour it fell with a 'ping' into a metal dish. Alfonso's candle clock is significantly more elaborate than this. Although Alfred Wegener opines, albeit tentatively, that Alfonso is the inventor of this mechanism,[34] the introduction to this treatise states:

> Otra manera fallamos de relogio que es muy buena. et muy conuenible pora poner eneste libro. & dizen a este estrumente; el relogio de la candela. & pueden saber por el las horas passadas del dia. o de la noche: & ell ascendente. & el medio cielo. Et ardiendo la candela desde la prima noche hata la mannana; a de parescer della siempre fuera dela for-

ma cosa sennalada; ni mas; ni menos. Et por que entendiemos que era cosa apuesta & con pro. Mandamos a Samuel el Leui de Toledo nuestro iudjo que fiziesse este libro en que fabla de cuemo se deue fazer este relogio. & de cuemo deuen obrar con el.[35]

[We found another kind of clock that is very good and very useful to be placed in this book. They call this instrument the candle clock and one can know through it the passing of hours of a day or night, the ascendant, and the zenith. And as the candle burns from the beginning of the night until the morning, one must see no more of the candle than the amount specified. And because we thought it to be elegant (*apuesta*) and advantageous (*con pro*) we ordered Samuel el Levi of Toledo, our Jew, to write a book on how one can make this clock and how to operate it.]

Whether "fallamos" [we found] connotes "came across" or "invented," one obtains the notion that he had before him an actual model of the clock, and on the basis of this model ordered a book to be written on its construction and use. It is known, too, that the Arabs knew candle clocks.[36]

This candle clock, like the mercury clock, is weight driven. The weights are attached to strings on either side of the candle. The other ends of the strings run through pulleys and ultimately attach to either side of a specially made base at the bottom of a candle. As the weights shove the candle, or actually 'pull' the candle upwards, the top of the candle meets firm resistance and remains motionless while only the bottom of the candle rises in proportion to the melt of the candle at the top. Successful escapement in this device depends on the rate of melt versus the stress originating in the pull of the attached weights. Matters of weight and counterbalance are further complicated when a string, attached at one end to a graduated panel containing the signs of the zodiac, is then run through another pulley and attached to the left cord connecting the candle and the weight. To control the burn rate, Alfonso requires that the wick be of cotton, that it be the cleanest possible, and that the thickness of the wick be of such manufacture that the candle cannot drip. He further requires that the candle be as round and smooth as possible; that the wick be weighed for the making of future candles for the device. Finally, since the calibration of the panel which rises cannot be completed until the candle is actually burning, two things are assumed: 1) some other device, perhaps an astrolabe or quadrant, will be used to determine when 15 degrees of each hour have elapsed so that the panel can be appropriately marked, and 2) once it is marked, that future candles used in the device will burn at the same rate as the one used for establishing the calibration.

All of these variables seem to regulate the entire mechanism, as A. A. Mills suggests for the mercury clock, to the realm of fantasy rather than of functional fact. So much is left to the expertise of the craftsman constructing the object.[37]

Although Arab sources for these three treatises seem likely given the nature of Alfonsine 'science,' and although no one has yet determined what these sources might be, the work of Donald Hill is of interest. In the mid- to late seventies, he edited two Arabic treatises. One is by three ninth-century brothers known as the Banu (or sons of) Musà bin Shakir. Their work is *Kitab al-hiyal*, or "Book of Ingenious Devices."[38] Many of the Banu Musà's devices are elaborations of the basic ideas contained in the works of Philo or Hero. Others are apparently innovations. Some of the Banu Musà devices listed by Hill are "concentric siphons," "a boiler that discharges hot water when cold water is poured in," and, more pertinent to our discussion, "vessels that are constantly refilled."[39]

The other treatise edited by Hill is Iben al-Razzes al-*Jazari's Kitab fi ma rifat al-hiyal al-handaslyya*, or "Book of Knowledge of Ingenious Mechanical Devices."[40] Al-Jazari lived over 300 years after the Banu Musà, and his work incorporates most of the significant devices and techniques of his predecessors whom he acknowledges. His 1206 date is contemporaneous with Alfonso's. Hill writes:

> Almost all the ideas and techniques usually thought of as typically Islamic which later reappeared in the West, can be found in al-Jazari. A direct transmission of his work to Europe cannot, however, be postulated. There is no known medieval translation of it into a European language but, in any case, it is probable that the most important transmissions occured before his book was composed. It is important to observe that al-Jazari records the assistance of a continuing tradition of machine technology from the time of Banu Musà to his own day. He prides himself on being a part of this tradition ... [41]

It would appear that Alfonso too fits into the same tradition. Unlike the Banu Musà whose work with hydraulics does not condern the moving or activation of biological or cosmographical automata on monumental clocks, that of al-Jazari and the other Arabic writers does, as does that of Alfonso.

Although it is still unclear how Islamic ideas were transmitted to Europe, an obvious connection is the Iberian Peninsula. Two obvious points of contact are the Toledo School of Translators and the Scriptorium of Alfonso X, el Sabio. The notion of contact between al-Jazari's work and that of Alfonso is possible and tantalizing, but must await further scrutiny by Arabists. That Alfonso continues and fits into the tradition of "ingeneous devices," however, cannot be doubted.

We do have in *Libro del saber* a bona fide, medieval do-it-yourself manual. One should not, however, take the qualification "medieval" lightly. Unlike the popular do-it-yourself manuals of today, such as the *'X'-for-Dummies* series, in which everything is spelled out and illustrated, Alfonso's text demands considerable technological expertise from the reader, despite its claims for universal understanding, i.e. that the book was written to be understood by any man who read it.[42] The elaborate illustrations in the text, the superb filligree work seen throughout, the alternate use of red and blue initials, the display script in this codex, destined it, and all the Royal Scriptorium codices, for that matter, for inclusion in the library of the King. The codex itself was meant to be a showpiece of the scholarly patron's largesse and erudition.

One can hardly refrain from wondering whether the three gravity-powered clocks at least were also more ostentation than functional contraptions. Alfonso keeps the mechanism of each decidedly away from public gaze leaving only the resulting motion of the panels depicting the heavens, or in the case of the mercury clock, the 'rete' of the astrolabe -- a stereographic representation of the same -- , visible. In the treatise of the water clock, for example, we read:

> Et desi faz un cobertero a todel relogio & ala tinaia. de guisa que non paresca del ninguna cosa. si non la tabla quando suuier. non mas. & desta manera pareçra el relogio mas fremoso quando fueren ascondidos todos sus engennos.[43]
>
> [And so encase the entire mechanism of the clock so that no part of it is visible except the panel as it rises. And in this way the clock will seem more beautiful when its mechanism is hidden.]

Only the astrolabe of the mercury-clock is to be visible. The rest is placed behind a wall made of planks.

> Et quando esto quisieres fazer; faz una pared de tablas. & pon tras aquella pared; todas las ruedas sobredichas armadas assi cuemo te ya dixe. & priega la lamina dell astrolabio en la faz de la pared de las tablas. & saca el cabo de la piertega por el forado de la lamina dell astrolabio. & faz una sennal en la red enel cerco de los signos enel grado do es el Sol aquel dia. & suelta el relogio quando nasciere el Sol.

> & pon el grado del Sol sobrel orizon. & aprieta la piertega con la red. & lexa el relogio correr. & mouersa la red sobre la lamina del astrolabio poc a poco todol dia. & la sennal que feziste enel grado del Sol; te mostrara que tanta es la altura en qual hora quier del dia.[44]
>
> [And when you wish to do this, make a wall of boards and place behind that wall all the aforementioned wheels arranged the way I already told you. And nail the plate of the astrolabe on the face of the wall of boards. and pull the end of the shaft through the hole in the plate of the astrolabe; and make a sign on the rete on the circle of the signs in the degree where the sun is that day; and release the horologe when the sun rises; and put the degree of the sun over the horizon. and join the shaft with the rete and let the horo-loge run; and the rete will move over the plate of the astrolabe little by little the entire day; and the mark that you made in the degree of the sun will indicate to you what the altitude is at any hour of the day.]

The candle clock may have caused the greatest wonder of all. To the right one viewed a burning candle which never burned downward out of sight. At the same time to the left a graduated panel containing the signs of the zodiac which appeared at the same time as they became visible in the heavens rose into view:

> & quando esto ouieres fecho; auras armada la candela pora quemar. & accender la as quando el Sol se pone. & taia toda uia el pauil quanto uieres que sera grand por que non se uierta la cera. & por que no queme mas de su derecho. & esto faras una uez en cada hora o mas si lo ouiere menester segund la gordez del pauil. & assi commo se ua la candela quemando assi se ua subiendo deguisa que ninguno no la pueda ueer. & deguisa que pareçca siempre dell estado que la delexeste deprimero. & assi commo se quema la candela; assi se ua subiendo; otrossi la tabla de los signos.[45]
>
> [and when you have done this, you will have prepared the candle to burn. And light it when the sun sets, & cut the wick as much as you see that is large so that it not drip wax, and so that it not burn more than is correct. and you will do this one time in each hour or more if it is necessary according to the thickness of the wick. And thus, as candle continues to burn, it thus rises so that no one can see it, and so that it appear always to be in the state you left it in at first. And thus as the candle burns, so it goes rising, and also the panel containing the signs (of the zodiac).]

Alfonso's intention to awe can neither be denied nor ignored.

In closing then, let it be said that in spite of the difficulty of properly assessing the degree of originality of the mechanisms described within the clock unit, in spite of the difficulty of determining to what degree these mechanisms were fanciful and to what degree they were functional, there can be no doubt as to their ingenuity. None can be considered an ignoble construction. Perhaps Alfonso's clocks more than any of the other instruments determine to what degree Alfonso was in the technological fore, and provide another reason why the sobriquet "el Sabio" is well-merited. They can indeed be considered monuments to his patronage and to what he would want us to consider his creativity, an example of *homo faber* at his best.

University of New Mexico

Notes

1. See A(lan) D. Deyermond, *A Literary History of Spain: The Mi-ddle Ages* (London: Ernest Benn Limited; New York: Barnes & Noble, 1971), 87-95.

2. As is customary, my reference to Alfonso as author of his works is merely for convenience and in no way wishes to address what Alfonso himself actually wrote. I acquiesce to Evelyn S. Procter's summation that Alfonso was a dedicated editor patron in "The Scientific Works of the Court of Alfonso X of Castile: The King and his Collaborators," *Modern Language Review* 40 (1945), 12-29 at 12, and to Antonio G. Solalinde's unsurpassed study on the matter of Alfonsine authorship: "Intervencíon de Alfonso X en la redaccìon de sus obras," *Revista de filología espanola* 2 (1915), 283-8.

3. Alfonso el Sábio, *General estoria, segunda parte*, ed. Antonio G. Solalinde, Lloyd A. Kasten, and Victor B. Oelschläger, vol. 2 (Madrid: Consejo Superior de Investigaciones Científicas, 1961), 31. All trans-lations provided are the author's.

4. Francisco Rico, *Alfonso X el Sabio y la 'General Estoria': tres lecciones* (Barcelona: Ariel 2nd ed., rev. & enlgd. 1984), 115.

5. "Charles F. Fraker, Alfonso X, the Empire and the Primera crónica," *Bulletin of Hispanic Studies* 55 (1978), 95-102 at 95.

6. *Alfonso X, o Sabio, Cantigas de Santa Maria*, ed. Walter Mett-mann, vol. 1 (Vigo: Edicións Xerais de Galicia, 1981), 102, verses 19-26.

7. Alfonso el Sabio, *Libros de acedrex, dados e tablas*. Das Schachzabelbuch König Alfons des Weisen mit 51 Miniaturen auf Tafeln nach der Handschrift J. T. 6 fol. des Escorial mit Glossar und Gram-matischem (sic) Abriss Herausgegeben und Übersetzt von ..., ed. and trans. Arnald Steiger (Genève: Librairie E. Droz, and Zürich-Erlenbach: Eugen Rentsch Verlag, 1941)

8. Miguel Asín Palacios, "El juicio del P. Mariana sobre Alfonso el Sabio," *Al-Andalus* 7 (1942), 479.

9. Richard H. Allen, *Star Names: Their Lore and Meaning* (1st ed. 1899, New York: Dover, rev. rpt. 1963), 12.

10. Such attributions, of course, have given rise to any number of defenses, for example, see the exaggerated one of Manuel Rico y Sinobas, ed., *Libros del saber de astronomia*, vol.1 (Madrid, 1863), xvi who protests that Alonso could not in any way be interested in so spur-ious an endeavour as "astrology"; or see the very rational one of José Soriano Viguera, *Contribución al conocimiento do los trabajos astr-onómicos desarrollados en la escuela de Alfonso X el Sabio* (Madrid: Tipografía de Alberto Fontana, 1926), 26, who makes clear that such a declaration of hubris cannot be found in any of the writings of the Learned Monarch.

11. All of these purposes can be found in Alfonso's *Libro de las formas & ymagenes* in the chapter rubrics (identified by: page number, column ['a' or 'b'], and line number) as edited by Roderic C. Diman and Lynn W. Winget (eds.), *Alfonso El Sabio. 'Lapidario' and 'Libro de las formas & ymagenes'* (Madison: Hispanic Seminary of Medieval Studies, 1980). I do not reproduce their use of italics to indicate their expansion of scribal abbreviations: "El .viij. pora fazer guerras & corrimento de sangres"(164a83); "El .xxij. pora auer nombradia de sabio. & non le sea grieue ningun saber" (153b87); "El .iii. pora seer el rey temido & obedecido de todos los otros reyes; & de todos los omnes" (154a64-5); "El .xv. pora que al rey o al princep le quiera bien su pueblo; & quel amen" (156b22-3); "El .xxvi. pora fazer foyr las chinches & las pulgas" (157a95).

12. Jean Leclercq, "The Experience of Time and Its Interpretation in the Late Middle Ages," in John R. Sommerfeldt and E. Rozanne Elder (eds.), *Studies in Medieval Culture VII and IX* (Kalamazoo: The Medieval Institute, Western Michigan University, 1976), 137-50 writes: "It was for the divine office that an attempt was made to measure time by observation of stellar movement, ..." (138).

13. See Anthony J. Cárdenas, "A New Title for the Alfonsine Omnibus on Astronomical Instruments," *La Corónica* 8 (1979-80), 172-8.

14. The works as named in MS 156 are: 1) "libro de las .xlviij. figuras de la .viija. espera" [book of the 48 constellations in the 8th sphere; star catalogue]; 2) "espera redonda" [celestial globe]; 3) "armellas del ataçyr" [armillary of the athazer]; 4) astrolabio redondo" [spherical astrolabe]; 5) "astrolabio llano" [flat astrolabe]; 6) "lamina universal" [universal plate]; 7) "açafeha" [saphea]; 8) "armellas" [armillary sphere]; 9) "laminas de las siete planetas" [plates of the seven planets]; 10) "quadrante" [cuadrant]; 11) "piedra de la sombra" [sundial]; 12) relogio del agua [clepsydra]; 13) "relogio de argent vivo" [quicksilver hologe]; "relogio de la candela" [candle clock]; 15) "palacio de las horas" [palace of the hours]; 16) "ataçyr llano" [flat athazer]. According to their prologues, three of these had been produced originally in the 1250s: 1256 (no. 1), 1259 (no. 2), and 1256 (no.7).

15. For a study of this instrument see Mercè Viladrich and Ramon [sic] Martí, "Sobre el Libro dell ataçir de los Libros del saber de astronomia de Alfonso X el Sabio," in *Nuevos estudios sobre astronomía espanola en el siglo de Alfonso X*, ed. Juan Vernet (Barcelona: Instituto de Filología, Institución

'Milá y Fontanals,' Consejo Superior de Investigaciones Científicas, 1983), 75-100.

16. The text of the sundial begins with a diagram that ends chapter 3. Thus, the first full chapter is chapter 4 -- "De saber la altura del sol por las horas passadas del dia" ["On knowing the altitude of the sun by means of the hours of the day which have passed"]. The last treatise, "palacio de las horas" [palace of the hours] ends a few lines into chapter 4 of the second part of the treatise.

The text of the *Libro de saber* in my transcription of MS 156. In it I follow the text verbatim and do not indicate expansion of scribal abbreviations, nor add punctuation; otherwise the text is as found in this codex. I indicate the text location in the notes and elsewhere by reference to foliation plus 'r' (= recto) or 'v' (= verso).

17. F.M. Feldhaus, "Die Uhren des Konigs Alfons X, von Spanien," *Deutsche Uhrmacher-Zeitung* 44 (1930), 608-12. Despite the promising title of the article and of the journal itself, this study provides only the most general of descriptions and does not treat our topic.

18. J. Eric Haswell, Horology: *The Science of Time Measurement and the Construction of Clocks, Watches, and Chronometers* (London: Chapman and Hall, 1937), 44.

19. Ibid., 46.

20. Lynn White, Jr., *Medieval Technology and Social Change* (1962; rpt. Oxford: Oxford UP, 1970), 119.

21. Ibid., 119.

22. MS 156, *Libro del saber*, fol. 176r.

23. For example: Feldhaus, "Die Uhren," 608-12; José A. Sánchez Perez, *Alfonso el Sabio: Siglo XIII* (Madrid: Aguilar, 1935), 91-2, 102.

24. For example: Harry C. Brearley, *Time Telling Through the Ages* (New York: Doubleday, Page, & Co., 1919), Ch. 4 "Telling Time by the Water Thief" 49-58; G. H. Baillie et. al., *Britten's Old Clocks and Watches and Their Makers* (New York: Bonanza Books, 7th ed. 1956), 3; or Morris Hicky Morgan (trans.), *Vitruvius: The Ten Books on Architecture* (1914; rpt. New York: Dover, 1960), Ch. 8 "Sundials and Waterclocks" 273-7.

25. Vitruvius records an alternative, albeit a more expensive one, when he claims that Ctesibus made "an orifice in a piece of gold, or by perforating a gem, because the substances are not worn by the action of water, and do not collect dirt so as to get stopped up" (Morgan, *Vitruvius*, 174).

26. H. Alan Lloyd, *Some Outstanding Clocks Over Seven Hundred Years 1250-1950* (London: Hill, 1959), 4-5.

27. MS 156, *Libro del saber*, fol. 185r.

28. Alfred Wegener, "Die astronomischen Werke Alfons X," *Bibliotheca Mathematica* 3. serie, 6 (1905), 129-85 at 163.

29. Lloyd, op. cit. 4-5.

30. For a discussion on the relationship between the plane astrolabe and the anaphoric clock see A.G. Drachmann, "The Plane Astrolabe and the Anaphoric Clock," *Centaurus* 3 (1954), 183-9 and for the relation-ship between

the modern clock and the anaphoric clock see Derek J. Price, "The Prehistory of the Clock," *Discovery* (April 1956), 153-7 ff.

31. A. A. Mills, "The Mercury Clock of the Libros del Saber," *Annals of Science* 45 (1988), 329-44.

32. For a discussion of this problem see Georg Bossong, "Las traducciones alfonsíes y el desarrollo de la prosa científica castellana" in *Actas de Coloquio hispano-alemán Ramón Menéddez Pidal* (Madrid, 31 de marzo a 2 de abril de 1978), ed. Wido Hempel and Dietrich Brie-semeister (Tübingen: Niemeyer, 1982), 1-14.

33. White, *Medieval Technology*, 120-1.

34. Wegener, "Die astronomischen Werke," 163.

35. MS 156, *Libro del saber*, fol. 189r.

36. Donald R. Hill (trans. and annot.), *The Book of Ingeneous Devices (Kitab al-Hiyal) by the Banu (sons of) Musà bin Shakir* (Dordrecht: D. Reidel Publishing Co., 1979), 21.

37. See Hill, Ibid., 11 for a similar observation with regard to Fakhr al-Din Ridwan b. Muhammad al-Sa ati's (b. 1203) work.

38. Hill, Ibid.

39. Ibid., 21.

40. Donald R. Hill (trans. and annot.), *The Book of Knowledge of Ingeneous Mechanical Devices (Kitab fi ma rifat al-hiyal al-handasiyya) by Ibn al-Razzaz al-Jazari* (Dordrecht: D. Reidel, 1974.)

41. Hill, *Ingeneous Devices*, 22.

42. To begin, Alfonso X's intention, was, in his own words, to produce the most complete and perfect treatise ever written on the subject, such that anyone who used this text would have no need to consult any other:

> Et por que non fallamos libro en que fable de como se deue fazer de nueuo; Por end nos Rey don Alfonso el sobredicho. mandamos al icho Rabiçag que lo fiziesse bien complido & bien paladino. de guisa que lo entendiessen aquellos que ouiessen sabor de lo fazer nueua mientre assi como lo auemos fecho en los otros libros que fizemos delos otros estrumentes. (fol. 40r11-19).
>
> [And because we found no book that addressed how it should be made, thus we the King, the aforesaid don Alfonso, ordered Rabiçag to do it fully and clearly so that anyone, who desired to make it anew, could understand it just as we have done in the books on the other instruments.]

The following unmistakably elitist attitude, however, contradicts the repeated intention of making his book accessible to all through clarity:

> Et desto deuen seer percebudos los que quisieren obrar deste saber. Ca esta es una de las mayores poridades que y a. & que mas punnaron de encobrirlos sabios. Pero la su encubierta tenemos que la fizieron por que los omnes nescios nola sobiessen. ca pora los entendudos fizo dios el saber. (fol. 7v89-94).
>
> [Those who wish to work with this science should be perceptive because this is one of the greatest secrets that exists and that the sages

most tried to hide. But this desire to hide it, we hold, was so done that fools not know it. For it is for men with understanding that God made knowledge.]

How might this apparent contradiction -- knowledge for all vs. knowledge for the select few -- be resolved? It may be that Alfonso's democratic perspective was qualified by a good dosage of natural selectivity. His knowledge would be clear for those desirous of it providing they had the intellectual wherewithall to penetrate it. The next passage shows that even among the cognescenti there are different kinds of knowledge and not all are privy to all there is to know. Here Alfonso seems to refer to craftsmanship -- adding form to matter, as it were:

Et por que non son los mas delos entendudos maestres nin sabidores en parar las linnas & las fazes unas sobre otras. & en componer las & en guisar las; acaeceles que non son maestres ni sabidores en las sotilezas & en las primezas de las obras de las manos. Et por esso non pueden poner aquella forma que elloos toman & entienden por la sciencia; en materia niguna. ni saben como se puede auenir la materia con ella pora recebir la forma que ellos ymaginaran en sus entendimientos. Et por esto fazesseles grieue de lo sacar afecho. Et por tanto qiero mostrar como la puede fazer quien quier; por la mas ligera carrera. (fol. 133r.).

[And because the majority of the learned are not masters or knowers of how to prepare the lines and the faces, one over the other, nor of how to arrange them and to make them ready, it happens to them that they are neither masters nor knowers of the subtleties and of the principles of manufactured items. And for that reason they cannot place that form that they take and understand through science onto any material; nor do they know how to join material to it (form) so that form can receive what they imagined in their understanding. And thus it is difficult for them to produce it. And therefore I wish to show how anyone can make it and in the easiest way.]

43. MS 156, Libro de saber, fol. 182r
44. MS 156, Libro de saber, fol. 189r.
45. Ibid., fol. 194r.

Time and Travel in Old Norse Society

Thorsteinn Vilhjalmsson

1. Introduction

1.1 A Note on Context and Evidence

When a society produces intellectual works like the Eddas and the Icelandic sagas[1] you expect that this is not an isolated feat. Rather, you assume that such works have grown from a fertile soil which might have fostered other species of intellectual and cultural acheivements. Indeed, it might appear a matter of chance which type of works attracts attention.

The methods and views of history of science have been neglected in discussions of medieval Scandinavian society and culture. Both scholarly and popular accounts have overlooked the intellectual aspects of the transoceanic traffic initiated and maintained by the Vikings and their successors, perhaps because the literary achievements of the Old-Norse have outshone other aspects of their culture. Also, as there appears such a long way from more or less fictitious literature to practical navigation, many fail to see the cultural link.

Another reason is the diverse nature of our sources. In the case of the sagas we have most of the texts in a form close to the original. Since the text is the essence of the cultural achievement you may say that the sagas speak for themselves in the true meaning of the words, and you need no other evidence. On the other hand, in the case of the navigation, say, the evidence is more indirect and contextual. Thus, we have to make do with the general historical record, scarce and maybe indirect textual references to navigation, and archaeological data of limited geographical extension.

1.2 The Sources

The pieces of evidence we have concerning medieval Old Norse scientific knowledge and skill are of quite disparate types:

- Dedicated texts on astronomy, time reckoning and encyclopaedic matters date from around 1150-1400. They contain mainly common European material, although in the vernacular and adapted to the Northern reader.[2]

- Dedicated but short text inserts in various manuscripts of the imported scholarly texts report on independent observation and contemplation, for instance on calendar, solar motion and navigation.[3]
- Pedagogic texts educate the reader on various matters of practical value, as for instance the Norwegian *King's Mirror* from around 1260. Among other things it contains sections on astronomy, time reckoning, tides and related subjects for professional tradesmen.[4]
- The saga literature itself contains interesting information on the knowledge and ideas of the Old-Norse in the period from 850-1300. Recent scholarship is increasingly aware of this. It is a challenging project to sort out the knowledge and skills in question.[5]
- Ships and sailing gear have been excavated, mainly in Denmark and Norway, and replicas made from them for testing in practice. Thus, valuable data have been obtained on the seaworthiness of the ships, their approximate speed and so forth. In general terms these data tend to agree with the textual information.[6]
- Historical and archaeological facts like those of the more or less permanent Viking settlements overseas demonstrate the capability of maintaining regular ocean traffic.
- The recent excavation in L'Anse aux Meadows in Newfoundland proves that the Vikings were in the western hemisphere as stated in the saga material.[7] On the other hand, we may safely state that the presence of the Vikings in America was limited in space and time; they were at the verge of their capabilities.

This itemization shows that there is no single dominant comprehensive source of information. Hence, it is crucial to bring all these fields of information together in order to assemble a clear, broad and balanced picture.

1.3 Time and Calendar

All organized societies need to have various ways of handling time. On the one hand, you may discern the need for dating various actions, meetings, etc. For this purpose people use what we call a *calendar*. On the other hand, we see the need for measuring *time intervals* in various contexts.

The Viking society of medieval Scandinavia is no exception to this. It is possible to suumarize Viking treatments of time quite briefly. They are an interesting mixture of indigenous methods and ideas on the one hand and specific items from the continental common goods on the other. The indigenous features appear to be grounded in their specific environment and culture. Thus they constitute an interesting object for a partly theoretical study of cultural history.

Almost everywhere on earth the climate entails *seasons* that are relevant to human activities. Therefore the usefulness of the calendar is universal albeit somewhat variable. For instance, the specific seasonal activities in question may vary. The relative importance of the calendar and its precision may hence also vary from one society to another, even at a similar stage of technological development. Thus, in farming societies, whether they be based on raising corn or cattle, grass or sheep, a good calendar is obviously quite useful. It helps the farmer to know when to should sow the grain and let the ram to the ewes, just to mention two examples of annually periodic activities which would otherwise be difficult to date in an optimal way.

As I have recently written an essay in English on the calendar in Icelandic and Old Norse society before literacy,[8] this will not be treated at any length here. Instead, this will focus on the treatment of time intervals in the Icelandic and the Old Norse society, a kind of time measurement is especially relevant in the context of travel.

1.4 Travel and Navigation

Given the nature of ancient or early medieval coastal navigation, people occasionally lost their way even in coastal travel. In some cases they might find new hitherto unknown lands and, what is most important for the historian, return to tell the story. Some of the lands thus found and reported tended to become lost again. This holds for the famous voyage of Pytheas of Marsilia to *Thule*, which might have been Iceland or a similarly out-of-the-way place in the North-Atlantic. In such cases people have at the time of "discovery" lacked the ability or the motivation to maintain regular traffic to the new land.

However, the Old Norse Vikings were not only able to find their way over the ocean on single, more or less stray voyages, but they also maintained regular transoceanic traffic for centuries. This was indeed necessary for their more or less permanent settlements in the British Isles, the Faeroes, Iceland and Greenland. To do this, they needed knowledge of and skills in what we now would classify as astronomy and other natural sciences. In the beginning, this knowledge was imported from the common goods of early medieval Europe but was subsequently nurtured and developed by independent observation and contemplation. This is explicitly documented in twelfth-fourteenth century writings in the vernacular on astronomy and calendar and other encyclopaedic matters.[9] Some of these specialized texts may be likened to modern monographs while others are more like textbooks. The documentation in the more celebrated saga literature is more implicit but not less interesting to decipher and interpret. The general thrust of their message as to knowledge and skills in the society fits quite well with the more direct evidence of the monographs. Modern archaeology is also providing strongly significant data on ships and sailing gear, and there is also recently discovered evidence of Old Norse presence in Newfoundland.[10]

Literacy and literary interest gradually entered Iceland and Norway from around 1100 onwards. It then became easier to obtain knowledge related to navigation from imported books rather than from first-hand observations or independent contemplation. At the same time, the dissemination of the best of ancient knowledge greatly improved the opportunity to acquire knowledge of astronomy and related matters from books. Consequently, the tendency towards originality seems to have diminished.

In practice, the Nordic Vikings expanded their sphere of activity on land and sea, reaching in the end the Black Sea, the North American continent, Sicily and the Volga River. On the seaside to the West they started with Shetland, the Orkneys, the Hebrides and the larger British Isles. After they had settled in the Faeroes, it was only a question of time when they would take the step to Iceland, in so far as they would consider it habitable. For that they would, among other things, have to be able to maintain regular sea traffic to this distant island. The knowledge and skill necessary for that had been developing and accumulating in the eighth and the ninth centuries, until Iceland finally was settled around year 870 AD, followed by Greenland about a century later. But in the case of Wineland they had stretched themselves beyond the limit of their capability to maintain a continuous link of traffic.

2. The northern sun and the time

2.1 The Peculiarities of Subarctic Solar Motion

One might think that solar motion would obey the same general principles when observed from arctic or subarctic regions as it does in the more temperate zones of the world. Actually, however, the solar motion in the North is quite different from the more temperate zones. We are not only concerned with the well known and prominent phenomena of the arctic midnight sun, the bright nights of the subarctic or the less frequently mentioned counterparts of these phenomena in the dark part of the year. Rather, we want to consider the differences which manifest themselves, for instance, in the possibilities of using the solar motion as a more or less accurate measure of time during the day.

Of course, the length of the diurnal arc of the sun can be used as a measure of time from any part of the earth. You would just have to be able to mark and memorize the places of sunrise and sunset on the horizon for the day of observation. However, this only works if you are staying at a fixed place and you have some prominent marks on the horizon or you take the precaution of making them yourself. In southern regions there may also be practical difficulties in measuring the arc when the sun is high in the sky, unless you have some kind of specialized instruments.

Here in the North this situation is quite different. The celestial pole is high in the sky and the diurnal orbits of the sun and the moon do not rise nearly as steeply as they do in the South. This means, for instance, that they can naturally be referred to the horizon almost all the time. The simple approximation of taking the speed of solar or lunar motion along the horizon to be constant is much better up here than it would be in more southern zones. This is reflected in several well-known peculiarities of speech in describing time and solar motion, both in medieval sources and also in rural language, place names etc., even today in Iceland.

2.2 The Octant Marks

A special feature of this kind in the Old Norse and Icelandic language is the concept of *eykt* in the sense of an eighth of the 24-hour day. The word "eykt" may be etymologically related to the number eight, a point to which we will return later. Hence, we may have an extra reason for picking the natural English translation of *octant*. The octants of day and night have special names as shown in table 1 (*vide infra*).

Corresponding to these eight time points of the day, Old Norse used with the term *átt/ætt* in the meaning of an octant of the circle or, in particular, of the horizon.[11]

Even now, in Icelandic farms so called *eyktamörk* or *octant marks* are quite common. These are normally some prominent landscape features on the horizon, as seen from the area in front of the farm houses ("hlað"). Thus you may have "Miðmorgunshnjúkur" (Mid-morning Peak), "Dagmálahóll" (Day Mark Hill), "Hádegisskarð" (Noon Pass) etc. Note that these names are heavily dependent of the farm in question; what is seen as a "Mid-morning Peak" from one farm would be a "Mid-evening Peak" for another farm on the opposite side of the peak.

Table 1. Names of the octants.

Local solar time	Term	Translation	Time of day	Term	Translation
0 a.m	miðnætti	midnight	12 noon	hádegi	high day
3 a.m.	ótta[12]		3 p.m.	nón, udorn, eykt	
6 a.m.	miður morgunn	mid-morning	6 p.m.	mið-aftann	mid-evening
9 a.m.	dagmál	"day mark"	9 p.m.	náttmál	"night mark"

The fundamental idea is that the sun should be above the horizon point in question at the time indicated by the name. Thus the sun should be above "Noon Peak" at 12 o'clock local solar time. And, sure enough, if this holds for a given day of the year it will be so all through the year, given that the peak is not so high as to hide the sun behind it during the darkest period of the winter.

However, when you come to the other octant marks, this is not quite so straightforward. You may, for example, pick your mid-morning octant mark to be straight east from your farm. Then you have ensured that the sun will be exactly at this mark at 6 a.m. local solar time at the vernal and the autumnal equinoxes, when it will be rising at this point of the horizon. But for the time between these two dates this will not be exactly so. For example, at the summer solstices in Northern Iceland, the sun will be some 10-15 degrees south of east at 6 a.m., corresponding to almost an hour's time. In other words, the "solar clock" obtained by using such octant marks might be up to an hour off in this region. (Needless to say, in more southerly regions this "clock" would be even more off, which presumably is one reason why it has not been used).

Modern people may find it difficult to visualize what would have been important here and what not. The methods described were introduced at a stage of technological development when people had no good independent means of checking the time. The error or uncertainty in time measurements implied by the subtleties in solar motion might very well have been tolerable, and even pass unnoticed at that stage. Even if people might have managed to spot the error by the means of a time-glass or some kind of a clepsydra, they might have found it too small to be of any concern for their purposes.

In fact, the annual rhythm of agricultural work in northern regions like in Iceland is such that time keeping through the day would have been most important during the summer, at the time of grazing, haymaking, etc. It is at least conceivable that people could have adapted their octant marks to this fact, picking them such that they would be "right" or, rather, consistent during summer at the cost of being further off during winter when the sun might not even pass them at all. To my knowledge, this has not yet been subject to scholarly study.

2.3 The Textual Sources

According to the sagas, the first Old Norse people to see America beyond Greenland were Bjarni Herjólfsson and his crew. According to the *Grænlendinga saga* (Saga of the Greenlanders) he intended to sail from Iceland to Greenland:

> Þá mælti Bjarni: "Óviturleg mun þykja vor ferð þar sem engi vor hefir komið í Grænlandshaf."
>
> En þó halda þeir nú í haf þegar þeir voru búnir og sigldu þrjá daga þar til er landið var vatnað en þá tók af byrina og lagði á norrænur og þokur og vissu þeir eigi hvert að þeir fóru og skipti það mörgum dægrum.
>
> Eftir það sáu þeir sól og máttu þá deila áttir, vinda nú segl og sigla þetta dægur áður þeir sáu land.[13]
>
> [Then Bjarni said, "This voyage of ours will be considered foolhardy, for not one of us has ever sailed the Greenland Sea."
>
> However, they put to sea as soon as they were ready and sailed for three days until land was lost to sight below the horizon. Then the fair wind failed and northerly winds and fog set in, and for many days they had no idea what their course was. After that they saw the sun again and were able to get their bearings; they hoisted sail and after a day's sailing they sighted land.][14]

What is most interesting here is the statement that when they saw the sun they could "get their bearings". This demonstrates quite clearly the strong connection between the sun and the directions in the minds of the people we are considering. This was around year 1000, so they had no exact instruments to speak of. They were at open sea and had lost all sense of orientation. So, therc is good reason to ask what they could use the sun for; what is "að deila áttir."

Of course, they could observe the culmination of the sun to tell them true south. But especially if this passage is compared with other sources, there is more than that. They could also observe or measure roughly the altitude of the sun at noon to tell them the latitude. This would not have been trivial since the noon altitude of the sun not only varies with latitude but also with the time of year. But if you had some kind of a formula or algorithm to cope with this variation you would be able to obtain a fairly accurate result. From the so-called *Oddi's Tale* from the middle of the twelfth century we have reason to think that the Old Norse people had a good and efficient formula or algorithm for this purpose.[15]

An interesting piece of text with a similar phrasing is found in the Norwegian *King's Mirror* from around 1260. After a description of the moon and the tides the author comments:

> En þessa hluti mega kaupmenn varla markað fá fyrir sakir svo skjótrar rásar því að tungl stígur svo stórum annaðtveggja upp eða niður að varla fá menn ættum skipað af rásum þess fyrir þá sök. En sól er fyrir því markandi að hún fyllir seinna rásir sínar hvorttveggja til uppstigningar og niðurstigningar og mega menn vel marka allar ættir af hennar rásum.[16]
>
> [But these things tradesmen can hardly mark for the sake of their swift course because the moon takes so large steps either up or down that people can for that reason hardly determine directions from its courses. But the sun can be marked because it fills its courses more slowly both in ascending and descending and people can well determine all directions from its courses.][17]

This text strongly supports the idea that people in this society knew how to cope with what we call the changes in declination of the sun for the purpose of determining geographical latitude. On the other hand, these changes in the case of the moon are so quick that it would have been much more difficult to use the moon for the same purpose.

So much for the "átt" for the moment. As to the time division of "eykt/octant"; the saga literature abounds with references to the individual octant divisions of the day as described above. When we would have stated the hour of an event the sagas simply state the octant instead, sometimes using, e.g., half an octant when such a precision is needed.

When we come to the relation of the octant marks and the sun the most famous quotation is again from the *Grænlendinga saga*. When describing the circumstances in "Vínland" (Wineland or The Land of Vine, i.e. some place on the eastern coast of North America) one of the observations goes as follows:

> Meira var þar jafndægri en á Grænlandi eða Íslandi. Sól hafði þar eyktarstað og dagmálsstað um skammdegi.[18]
>
> [In this country, night and day were of more even length than in either Greenland or Iceland. The sun passed the points of "eykt" and "day mark" in the period of the shortest days.][19]

In modern language the most straightforward interpretation of this implies that the sun would pass the horizon points of Southeast and Southwest[20] in the shortest day of the year. You may also note that the author clearly finds this worthwhile to report.

The main purpose of bringing this up here is that of showing the reader how the Old Norse seafaring people used available means for orienting themselves and gaining information on their location and their surroundings. It is therefore not of major concern there what exactly this tells us about the actual location. Let it suffice to say that this has been a matter of strong controversy and that, in my opinion, it must imply that the place in question *must* have been south of 58 degrees north, but it *could* have been much further south.[21]

2.4 The Question of Instruments

The abundance of all kinds of tools and instruments in modern society is such that we tend to find it difficult to do anything without such aids. Thus, looking at the practical navigational skills of the Old Norse, we find it incumbent to ask about the instruments they used for finding their way over the ocean, keeping time etc. Although they had no exact "hardware" instruments per se, they might rather have had useful "software" equipment in the form of knowledge of winds and currents, sun and stars, etc.

This question may strongly depend on what is called an "instrument". If somebody holds out a stick to make a rough measurement of angles, would you call that an instrument? What if one uses fist or fingers instead? Clearly, you would not expect to find evidence of methods of this kind through archaeology or even through direct textual reference. So, your only way to make inferences would be through observing what the people accomplished and thinking about how they did it.

A well-known candidate for a Viking age navigational "instrument" has been called a *bearing-dial*. It is a disk-shaped piece of wood with a notched edge found in an excavation near Narsarsuaq in Greenland in 1948.[22] The notches on the edge of the disk are fairly regular although not overwhelmingly so. Some authors have also wanted to lend significance to some scrapes on the surface of the disk, which to other observers will look like incidental scratches. I would think that such a device might have been used for time-keeping, e.g. at a farm. The peculiar northern feature of the nearly horizontal motion of the sun through the day, as described above, would have made this simple device useful in certain circumstances, e.g. for people living at a southern shore with no natural direction marks near the route of the sun. This might also hold for people staying through the day at a given place in otherwise unknown territory. But for people at open sea, it is not easy to see how such a rough device would add decisive and significant information to what they would be able to get without it.

Another puzzling candidate for the role of a navigational instrument is the so-called *solar stone*.[23] It has been suggested that this was one out of several possible minerals now known to be sensitive to the polarization of light. Several such minerals occur naturally within the sphere of activity of the Old Norse. And, for sure, such a polarizing stone could have been used to find the direction of the sun in certain circumstances similar to the context of the main medieval textual report on a solar stone, which we will now relate.

The source is the so-called *Rauðúlfs þáttur,* a kind of a short story preserved in an early fourteenth century manuscript but perhaps composed century earlier. The Norwegian king Ólafur helgi (Saint Olaf, d. 1030) is visiting a rich and wise farmer, Rauðúlfur. Late in the night one of his sons, Sigurður, is asked about his capabilities and replies that he has mastered one skill:

> "Það er að greina gang himintungla þeirra er eg sé og kenna stjörnur þær er stundir merkja svo að eg mun vita lengd um dag og nótt þó að eg sjái eigi himintungl og veit eg þó grein allar stunda bæði dag og nótt." Konungur svarar: "Þetta er mikil íþrótt."[24]
>
> ["That is to discern the motion of heavenly bodies, those which I see, and to know the stars which mark the hours, so that I will know time length at day and night although I do not see the celestial bodies, and still I know how to discern all hours both day and night." The king replies: "This is a great skill."][25]

The next day, King Ólafur conducts a test:

> Veður var þykkt og drífanda sem Sigurður hafði sagt. Þá lét konungur kalla til sín Sigurð og Dag. Síðan lét konungur sjá út og sá hvergi himin skýlausan. Þá bað hann Sigurð segja hvar sól mundi þá komin. Hann kvað glöggt á. Þá lét konungur taka sólarstein og hélt upp og sá hann hvar geislaði úr steininum og markaði svo beint til sem Sigurður hafði sagt.[26]
>
> [The weather was thick and snowy as Sigurður had predicted. Then the king summoned Sigurður and Dagur [Rauðúlfur's sons] to him. The king made people look out and they could nowhere see a clear sky. Then he asked Sigurður to tell where the sun was at that time. He gave a clear assertion. Then the king made them fetch the solar stone and held it up and saw where light radiated from the stone and thus directly verified Sigurður's prediction.][27]

This account is not too far from the possible, although somewhat exaggerated. Thus, the weather should not have been too "thick" since a part of the sky around zenith would have to be visible. Also, it seems somewhat besides the point to talk about light radiating from the stone. But the main point concerns the usefulness of a "solar stone" with the polarising properties mentioned above. Such a stone would actually have been quite helpful to a farmer for following the solar motion along the horizon during a partly overcast day. By comparison with knowledge gained from clear days, he would have been able to translate this to an observation of time.

This is the only Old Norse text referring to this kind of stone in some detail. You may note that the text describes its use *on land* only. On the other hand, it is not at all obvious how this could have been used as an efficient navigational instrument yielding significantly more information than could otherwise have been available.[28]

The compass was not introduced in Europe until around year 1200, and extant texts refer to some primitive versions of it in Scandinavia in the thirteenth century.[29] Thus I think we can safely conclude that in the first centuries of Old Norse ocean traffic people had no advanced observational or navigational instruments worthy of the name.

3. Time in Travel and Navigation

3.1The Report of Ohthere

When king Alfred the Great had his famous Orosius translation made in the last decade of the ninth century, some other texts were included. Among those was the report of a certain Ohthere from Hålogaland (Helgoland) in Norway who had sailed to England and visited the court:

> Ohthere sæde ..., þæt he ealra Norðmonna norþmest bude ...
>
> He sæde þæt he æt sumum cirre wolde fandian hu longe þæt land norþryhte læge, oþþe hwæðer ænig mon be norðan þæm westenne bude. Þa for he norþryhte be þæm lande; let him ealne weg þæt weste land on ðæt steorbord, 7 þa widsæ on ðæt bæcbord þrie dagas. Þa wæs he swa feor norþ swa þa hwælhuntan firrest faraþ. Þa for he þa giet norþryhte swa feor swa he meahte on þæm oþrum þrim dagum gesiglan. Þa beag þæt land þær eastryhte, oþþe seo sæ in on ðæt lond, he nysse hwæðer, buton he wisse ðæt he ðær bad westanwindes 7 hwon norþan, 7 siglde ða east be lande swa swa he meahte on feower dagum gesiglan. Þa sceolde he ðær bidan ryhtnorþanwindes, for ðæm þæt land beag þær suþryhte, oþþe seo sæ in on ðæt land, he nysse hwæþer. Þa siglde he þonan suðryhte be lande swa swa he meahte on fif dagum gesiglan. Ða læg þær an micel ea up on þæt land. ... ac him wæs ealne weg weste land on þæt steorbord, butan fiscerum 7 fugelerum 7 huntum, 7 þæt wæron eall Finnas; 7 him wæs a widsæ on ðæt bæcbord.[30]
>
> [Ohthere told ... that he dwelt northmost of all the Northmen. ... He said that he was desirous to try, once on a time, how far that country extended due north, or wether any one lived to the north of the waste. He then went due north along the country, leaving all the way the waste land on the right, and the wide sea on the left, for three days: he was as far north as the whale-hunters go at the farthest. Then he proceeded in his course due north, as far as he could sail within another three days; then the land there inclined due east, or the sea into the land, he knew not which, but he knew that he there waited for a west wind, or a little north, and sailed thence eastward along that land as far as he could sail in four days; then he had to wait for a due north wind, because the land there inclined due south, or the sea in on that land, he knew not which; he then sailed thence along the coast due south, as far as he could sail in five days. There lay a great river up in that land ... but all the way he had waste land on

> his right, except fishermen, fowlers, and hunters, all of whom were Fins, and he had constantly a wide sea to the left.][31]

First of all, we may realize that this is roughly a good description of the coast of Norway and then the Kola Peninsula and the White Sea. It is quite clear that it is based on real experience of the route. However, when we look at the details we soon recognize that the coastline in question is not so rectangular in shape as suggested by the bearings north-east-south. Thus, there does not seem to be a firm base for a detailed specification of the three sections of 2, 3, 4 and 5 days, respectively. Still, you may presume that the voyage started at Ohthere's homestead in Halogaland, and ended at the present location of Arkangelsk as suggested by the description of the river (Dvina). You then get an average sailing speed of around 3 knots.

Some authors seem to think that this speed is somehow too small as an average for sailing day and night.[32] Hence they speculate that Ohthere only sailed by day and took to land for the night. Though Norwegian skerries are not easy to handle in darkness, Ohthere did have bright nights. Thus daily landfalls on an unknown coast might have been riskier than just sailing at "night".

Thus we find the average effective speed of 3 knots to be a reasonable figure which in turn reinforces the trustworthiness of the report. At the same time, we may notice that the whole story is quite down-to-earth and contains an interesting emphasis on the time intervals as an important navigational item in describing the route.

3.2 Reports from the Saga Literature in General

The best known and the most prominent part of the medieval Old Norse literature consists in the so called *sagas of Icelanders* or *Íslendinga sögur*.[33] They were written in Iceland in the thirteenth and fourteenth centuries and each of them is normally preserved in several manuscripts, albeit not originals. They report on events that may have happened roughly in the period from 930 to 1030. They do this mostly in a secular, down-to-earth way with not too many supernatural phenomena entering the narration.

Since the literary quality of the Icelanders' sagas is such that they may be read as novels, the scholarly view of them as historical sources has vacillated. Recent scholarship has tended to read them as valid historical sources in a general sense albeit not in detail. It may then depend on the context whether you conceive them as describing the period of the writing or the earlier period which they purport to describe.

The sagas describe numerous summer voyages. Most start or end in Iceland although there are also long voyages between other countries. In many cases the time span of the voyage is mentioned, typically a couple of weeks, say, for a voyage between Iceland and Norway which was the most frequented route.The seamen frequently experience overcast or rainy weather so that they are unable to orient or locate themselves or "get their bearings" as it is called in the quotation on Bjarni Herjólfsson above. In these more unfortunate cases the voyage may take months and it is not at all certain that the ship finally reaches the land that the seafarers had in mind. This is quite apart from the voyages that ended in shipwreck, either at open sea or on some coast that might have been unknown and desolate; but we do not have direct reports of such voyages.

From the sagas and other parts of the medieval Old Norse literature it is clear that people gradually accumulated all kinds of knowledge relevant to sailing in the North Atlantic. This somewhat specialized knowledge would reside with a well-defined group of people -- the tradesmen or the professional navigators. One of the most efficient methods at their disposal was that of *dead reckoning* as it is called by English-speaking seamen.[34] It consists in following closely the time span of the voyage as a whole and of single stretches, for example how long you sail in a given direction. The ocean sailors have taken this custom from their forerunners, the coastal navigators and even the people doing long-distance land travelling for whom this was a well known and useful tradition. The Old Norse use of the method emerges clearly from marine itineraries that have probably been consciously and purposely utilized in navigation. As an example we give the following one from the Icelandic *Book of Settlements*:

> Svo segja vitrir menn, at úr Noregi frá Staði sé sjö dægra sigling til Horns á austanverðu Íslandi, en frá Snæfellsnesi fjögurra dægra sigling til Hvarfs á Græn-landi. Af Hernum af Noregi skal sigla jafnan í vestur til Hvarfs á Grænlandi, og er þá siglt fyrir norðan Hjaltland, svo að því að eins sjái það, að allgóð sé sjávar sýn, en fyrir sunnan Færeyjar, svo að sjór er í miðjum hlíðum en svo fyrir sunnan Ísland, að þeir hafi af fugl ok hval. Frá Reykjanesi á sunnanverðu Íslandi er þriggja dægra haf til Jölduhlaups á Írlandi í suður; en frá Langanesi á norðanverðu Íslandi er fjögurra dægra haf til Svalbarða norður í hafsbotn, en dægurs sigling er til óbyggða á Grænlandi úr Kolbeinsey norður.[35]
>
> So the wise men say, that from Stadlandet in Norway there is seven days' sailing to Horn in the East of Iceland, but from Snæfellsnes four days' sailing to Cape Farewell in Greenland.[36] From Hennøya in Nor-way you should head

> always to the west to Cape Farewell in Greenland, and then you sail to the north of Shetland, so that you only see it if the sight at sea be fairly good, but to the south of the Faeroes, so that the sea is midway in the mountain slopes, but in such a way to the South of Iceland that you observe birds and whales from there. From Reykjanes in the South of Iceland there is a sea of three[37] days to Slyne Head (?) in Ireland to the South; but from Langanes in the North of Iceland there is a sea of four days to Jan Mayen[38] in the North at the bottom of the sea, but a day's sailing to uninhabited land in Greenland from Kolbeinsey in the North.[39]

The most important and the best substantiated of these data is the first point of the trip from Norway to Iceland, a stretch that was very frequently sailed. The distance in question is about 530 nautical miles so that the average time of 7 days corresponds to 76 miles per day or 3.2 knots on average. Such an average implies that single voyages might well have been twice as fast, thus taking only, say, 3-4 days. All of this fits fairly well with information from other sources and also with modern experiments with replicas of Viking ships.

The other data in the text are more complex. However, taking the most plausible reading in each case, and taking the figure of "1 day" for the Kolbeinsey-Greenland stretch with the uncertainty implied by the figure,[40] the data give speeds in the range 3-6 knots. In several of these cases the data can only have been based on single stray voyages. They may have been undertaken involuntarily so that the likelihood is that they were done under favorable circumstances. Of course you are quicker in moving at sea when the wind more or less forces you to take a given route than you are when you are going to a certain destination and then put yourself at the mercy of the caprices of North Atlantic weather.

3.3 The Wineland Sagas

Two of the sagas of the Icelanders may reasonably be called "Wineland Sagas", the *Grœnlendinga saga* and *Eiríks saga rauða* (the *Saga of the Greenlanders* and the *Saga of Eirik the Red*). They are readily available in English in a thoroughly edited translation for Penguin by Magnus Magnusson and Hermann Pálsson.[41] They contain several reports of early eleventh century voyages of the Old Norse habitants of Iceland and especially Greenland to America. The first sighting was done by Bjarni Herjólfsson as already mentioned. The quotation above ends with the sighting of the first land and continues as follows:

> [Þeir] ræddu um með sér hvað landi þetta mun vera en Bjarni kveðst hyggja að það mundi eigi Grænland. ... Síðan sigla þeir tvö dægur áður þeir sáu land annað. ... settu framstafn frá landi og sigla í haf útsynningsbyr þrjú dægur og sáu þá land þriðja. ..., sigldu nú fjögur dægur. Þá sáu þeir land hið fjórða. ... Bjarni svarar: "Þetta er líkast því er mér er sagt frá Grænlandi og hér munum vér að landi halda." Svo gera þeir og taka land undir einhverju nesi að kveldi dags.[42]
>
> They discussed amongst themselves what country this might be. Bjarni said he thought it could not be Greenland. ... And after sailing for two days they sighted land once more. ... They turned the prow out to sea and sailed before a southwest wind for three days before they sighted a third land. ... They sailed now for four days, until they sighted a fourth land. ... "This tallies most closely with what I have been told about Greenland," replied Bjarni. "And here we shall go in to land." They did so, and made land as dusk was falling ...[43]

Upon closer exploration of the new lands sighted by Bjarni, his followers discerned three areas to the West and South of Greenland, called "Helluland" or "Slab Land", "Markland" or "Forest Land", and "Vínland" or "Wineland." The descriptions of the landscape, vegetation and the human inhabitants of the regions in question are quite realistic. There are several localities under the common denomination of "Wineland", the descriptions fitting roughly the area from the southern coast of the Gulf of St. Lawrence to Virginia.

The fact that the Norsemen were present in America in this period was corroborated beyond doubt in 1969 when a team led by the Norwegian scholars Anne Stine and Helge Ingstad had finished excavating a Viking site at L'Anse aux Meadows in Newfoundland.[44] Helge Ingstad thought that he had found Wineland itself but most scholars now agree that Wineland could not have been so far north and that the site should be seen as a station on the way further south.[45]

One fundamental question concerning time and the Wineland sagas is that of the feasibility of their travelling so far. The base of the Old Norse being in the North Atlantic, they would normally only have 4-5 months of the year for safe travel (roughly May to September for the case of Iceland, but beginning later in the case of Greenland because of sea ice). So there is real reason to ask: Would they have had time enough to get to Wineland in the summer in a reasonably regular way?

The Old Norse settlement in Greenland, "Eystribyggð" (Eastern Settlement), from which the Wineland explorers started their voyages was a little Northwest of the southern tip of Cape Farewell, near the present village of Narsarsuaq. They would normally begin by sailing several hundreds of miles to the North to obtain a shorter open sea passage to Baffinsland at the Davis Strait. This brings the total distance from "Eystribyggð" to "Wineland" up to the order of 2500 miles. Under favorable wind this would take some 20 days. In practice, however, you cannot expect to have such a wind for such a long time, so that a real one way trip is more likely to have taken 25-40 days.

Still, this can clearly be done in one summer, and if you are lucky, you can even do a round trip that is actually reported in some of the cases, although you shouldn't count on it.

Apart from the point that most of the Wineland voyages took about a half or a whole summer, the sagas do not contain too much information on sailing times. The most comprehensive report is that of the voyage of *Þorfinnur karlsefni* in Eirik's saga. Three times in this itinerary stretches of two days at open sea are mentioned but the duration of sailing along the coasts is not mentioned at all. This may reflect the fact that knowledge of sailing times is never so soothing as when you are crossing the open sea without the sight of land; it may help you in feeling that there is really something ahead. In coastal navigation this is not quite as crucial. The first stretch in question is clearly described as the passage of the Davis Strait between Greenland and Baffinsland, where the length also fits nicely. One of the other passages might be that of the Hudson Strait between Baffinsland and Labrador. It is not quite so clear where to look for the third open sea passage of this kind.

3.4 The Puzzle of the "dægur"

The length of the Old Norse voyages was such that "a day" would have been an adequate time unit for sailing reports. However, one of the characteristic features of early Icelandic or Old Norse time reckoning was its emphasis on the "dægur" instead of a "dagur" which for most purposes corresponds to "day" in English. In the context of sailing itineraries the meaning of the last word has been the subject of much scholarly scrutiny, speculation and controversy. Unfortunately some of the participants have lacked sufficient knowledge of the language in which this is embedded or of fundamental features of the natural environment in the North. This has given rise to even more controversy than would otherwise have been necessary.

In Old Norse as well as modern Icelandic the fundamental meaning of "dægur" is that of a common name for a day *or* a night. For comparison in English you may consider the use of "season" as a common name for summer or winter, or for summer, autumn, winter or spring. "Dægur" having no direct counterpart in English, it may often be translated as "half-day" although, in some contexts, such a translation may look like an awkward tautology. We are then referring to the word "day" in the somewhat astronomical meaning of the 24 hours' period of revolution of the earth around its axis. But we should note that in daily life "dægur" is not so strictly defined as to be a time unit of constant length, no more than a "day" in the everyday meaning of "the bright part of the diurnal 24 hour period".

The old Scandinavian word "dægur" later developed into "døgn/dygn" in Danish, Norwegian and Swedish where it attained the different meaning of the 24-hours day mentioned above. This seems to be one of the sources of scholarly confusion here, in spite of the fact that this change of meaning took place rather late. Thus, for instance the Norwegian author of the above-mentioned thirteenth century *King's Mirror* does not waver at all in his understanding of "dægur":

> En stundir má eg þér vel kunnar gera hversu langar þær kunna að vera því að 24 skulu vera á tveim dægrum, nótt og degi, meðan sól veltist um átta ættir, og verður svo að réttu tali að það eru þá þrjár stundir dags er sól veltist um eina ætt.[46]
>
> But I can easily acquaint you with the hours, how long they may be, for there shall be 24 in [two dægur] a night and a day while the sun rolls through eight directions [octants], and so it will be on the correct count, that there are three hours of the day when the sun revolves through one direction.[47]

In the manuscripts of *Rímbegla* or *Rím I*, a work on practical astronomy and time reckoning dating from around 1180 AD, the various time units are explained as follows:

> Ár heitir tvö misseri. Í misseri eru mál tvö, í máli eru mánuðir þrír, mánuði vikur vel svo fjórar í viku dagar sjö, í degi dægur tvö, í dægri stundir tólf.[48]
>
> A year is two semesters. In each semester there are two quarters, in a quarter three months, in a month a bit more than four weeks, in a week seven days, in a day two "dægur", in a "dægur" twelve hours.[49]

The emphasis on the term "misseri" is manifest in Old Norse texts in the frequent use of "tvö misseri/two semesters" instead of "one year".[50] The time reckoning system is also often referred to as "misseristal" (literally "counting of semesters"). The word "misseri" is actually a composite with an etymology corresponding directly to "half-year".[51] The English word "semester" is here used as a translation although in modern English it only has the academic meaning of "half of the academic year". According to standard dictionaries, its Latin and Greek origins correspond to "six months". The term "mál /quarter" is not as frequent or important as the others and will not be treated further here.

The week was of a special importance in the old calendar that might also be called "viknatal" (literally "counting of weeks"). From the period of the settlement around 870 AD to almost 1100, a year, or rather two semesters, would consist of an integer number of weeks, normally 52 but sometimes 53 so as to get the correct length of the average year.[52] With the "misseristal" and "viknatal" combined, people would give a date by a phrase like "fimmtudagur í tíundu viku sumars/Thursday in the tenth week of summer".

Returning to the "dægur", its relation to the day is the same as the relation of semester to the year. The use of "dægur" was also quite parallel to that of "misseri"; people would for instance speak of "two dægur" where we would say "one day".

So far, we can conclude that there is no doubt about the meaning of "dægur" as a time unit *in the context of pure and simple time-reckoning*. In such a context the word unambiguously would mean "12 hours".[53]

The "dægur" is of special importance when you turn to travelling, as you might indeed expect, since the time span of longer travel would be of the order of magnitude of several "dægur". We have already given several examples of its use in previous sections of the paper, in the context of itineraries and their time length. From a twelfth century itinerary by a pilgrim to Rome, we add the following one:

> Svo er sagt, að umhverfis Ísland sé sjö dægra sigling að rauðum byr, og skiptist svo sem þarf, því að eigi má eitt veður hafa. Svo og meðal Íslands og Noregs er kallað jafnlangt.[54]
>
> So they say, that it is a sailing of seven days around Iceland under a favorable[55] wind that should change as needed because you can not use the same wind all the time. So also between Iceland and Norway; it is considered equally far.[56]

The statement of a sailing of seven "dægur" between Iceland and Norway also appears in the quotation from the *Book of Settlements* given above. We there gave the average speed implicitly given by the various sailing itineraries to be of the order of 3-6 knots. In so doing, we tacitly assumed the word "dægur" *in this context of sailing distances* to mean "a day" or 24 hours. The most important point concerns the *speeds* which would be obtained by the other alternative of understanding "dægurs sigling" as "a sailing of 12 hours". This would bring the sailing speed of the various itineraries where "dægur" occurs up to the order of 6-12 knots. Thus we may describe the alternatives by the Table 2.

Table 2. Alternatives in the interpretation of "dægurs sigling".

Meaning of "dægurs sigling"	Number of hours of sailing	Average speed in itineraries
dagsigling = a day's sailing	24 hours	3-6 knots
half a day's sailing	12 hours	6-12 knots

First of all, we compare this with the following generalizing quote from the mid-thirteenth century *Rím II:*

> Á milli Björgvinjar og Niðaróss eru nær 4 gráður.[57] Þá verður ein gráða nær tylft sjávar. ... En jafn mikið er gráða á jörðu og tylft sjávar, en tvær tylftir ein dagsigling. Þá verður um kvaðrant jarðar 45 dagsiglingar, en í kringum jörð er meginhaf 180 dagsiglingar.[58]
>
> Between Bergen and Trondheim there are nearly 4 degrees. Then one degree is nearly a nautical dozen [i.e. a dozen of "vikur"]. ... But a degree on earth is the same as a nautical dozen, whereas two dozens are one day's sailing. Then the quadrant of the earth makes 45 day's sailings but around the earth along the grand ocean there are 180 day's sailings.[59]

This text is schematic and clearly emphasizing round figures in order to define "dagsigling/a day's sailing" as a kind of a unit of length. From the figures given you get the day's sailing as 2 degrees or 120 nautical miles, yielding an average speed of 5 knots. Thus, it fits somewhat better in with the first alternative described in Table 2, although the deviation from the speeds implied in the second alternative is not very decisive.

More decisive evidence is obtained from collecting and comparing all the reports on single voyages from the saga literature. One then sees that voyages between random places in Norway and Iceland most often take several weeks and that voyages of less than a week were extremely rare. It seems most natural to take the itineraries quoted above for this frequently trafficked stretch to be describing some kind of an average result of many voyages. Thus it would not fit in with the saga reports to assume the average time to have been 7 "dægur" in the literal sense of three and a half day.

A third piece of evidence comes from considering directly the capability of real ships. Several sources seem to agree that a ship like the so-called Viking ships (*knerrir*) used for Old Norse ocean traffic might make up to 12 knots in favorable wind, *relative to water.* Note that "favorable wind" here not only means "wind in a favorable direction" but also "wind of favorable force", i.e. not too slow and also *not too strong* since too strong a wind in the North Atlantic implies a rough sea which will force you to reduce your speed by lowering the sail. Anyhow, since the ship would only rarely be heading directly towards its destination this means that the *effective speed* (measured as the crow flies between the starting and ending points) would be somewhat (say 10-30%) less. In adverse wind, the effective speed would be much less, i.e. something like one or two knots only.

An average effective sailing speed of the order of 3-6 knots not only fits experiments with replicas of Viking ships. It also fits other data on sailing speeds in the history of navigation for the whole period from antiquity up to the late middle ages, where reports generally yield average speeds of the same order of magnitude.[60]

As mentioned above, an average effective speed of 3.2 knots on the route between Staður in Norway and Horn in Iceland, the shortest route between the two countries, will make the voyage last 7 days. A voyage between some other two places in each country would easily take twice as long as you may glean from a map. This actually fits nicely with the reports according to which voyages of several weeks were quite common and, for instance, a voyage of 4 days was considered quite exceptional.[61] Such a voyage corresponds to an average effective speed of 7 knots which is not implausible for single voyages. A normal effective sailing speed of the order of 3 knots in the North Atlantic also fits in with the experiences of modern sailboats in this area.

Now we have to ask, how it came about that "dægur" meant "12 hours" whereas "dægurs sigling" should mean "a sailing of 24 hours". Two observations may help to explain this translation of meaning.

First, there is the possibility of a so-called *pars pro toto* effect. When we give the age of domestic animals in Icelandic, we often state it in "winters" instead of years. Thus we may say of a 7 year old horse that it is "sjö vetra/seven winters old." Also, we state the age of the moon in "nights" rather than days even if we have a special word ("sólarhringur", corresponding to "døgn/dygn") to emphasize that we mean the day of 24 hours. This way of speech might have been used also in the case of the "dægurs sigling".

Another and maybe even more convincing point is that of the circumstances in sailing. This was mostly done during high summer where the night is bright so that the total diurnal period of 24 hours may look like one bright day or one "dægur".

We thus come to the conclusion that, on the one hand "dægur" in normal speech certainly means "half-day", but the composite "dægurs sigling" in Old Norse may in effect quite well mean "24 hours' sailing". As a minimal conclusion we may state that this must be the meaning of the words *if* we take the itineraries to describe some kind of an average of real sailing times.

Conclusions

The Old Norse society had its own ways of measuring and treating time. These methods were related to the natural environment of the subarctic and to the needs of society at this stage of technological development. Thus the rural agricultural society developed its own methods for following time during the day, especially during the economically decisive period of summer. This was done by utilizing the oblique diurnal motion of the sun in the North and defining the so called "eyktir" of 3 hours and the corresponding octant marks on the local horizon. The knowledge of solar motion related to this also enabled these northern people to utilize the sun for rough observations of their geographical latitude on voyages. Thus, for instance, shrewd observations on solar motion are reported in the Wineland sagas.

The Old Norse society had essentially no instruments for measuring time or for navigation. A disk of the kind found in an excavation in Greenland would only have been of limited help for these purposes. There is no clear-cut evidence for a solar stone having been used at sea. It could, however, have been used for local timekeeping in partly overcast weather.

Time keeping was quite important in Old Norse navigation and this is manifest in the many itineraries giving details of dead reckoning, for instance the average time length for various stretches. From the various data yielding sailing speeds we conclude that the average effective speeds of Viking ships in regular ocean traffic were of the order of 3-6 knots. In order to reach this conclusion I have argued for a definite interpretation of the word "dægur" when used in sailing itineraries.

University of Iceland

Notes

1. For a recent overview see Jónas Kristjánsson, *Eddas and Sagas: Iceland's Medieval Literature,* trans. Peter Foote (Reykjavík: Hið íslenska bókmentafélag, 2nd edn., 1992).

2. The most important texts of this kind are printed in the diplomatic edition: Natanael Beckman and Kristian Kålund (eds.), *Alfræði íslenzk: Islandsk encyklopædisk litteratur: II. Rímtöl,* (Copenhagen: Samfund til udgivelse af gammel nordisk litteratur, 1914-16).

3. Thorsteinn Vilhjálmsson, "Time-reckoning in Iceland before literacy", in Clive Ruggles (ed.), *Archaeoastronomy in the 1990s* (Loughborough, UK: Group D Publications, 1993), 72-4, and references therein.

4. The standard edition for reference is Ludvig Holm-Olsen (ed.), *Konungs skuggsiá* (Oslo: Kjeldeskriftfondet, 2nd. rev. edn., 1983). English translation in Laurence M. Larson (trans.), *The King's Mirror: (Speculum Regale - Konungs skuggsjá),* (New York: The American Scandinavian Foundation; London: Humphrey Milford; Oxford: Oxford University Press, 1917).

5. As to the sagas of the Icelanders, the recently published CD with the saga texts accessible for computer work and with a concordance will prove quite valuable for studies of this kind. See Eiríkur Rögnvaldsson et al. (eds.), *Íslendinga sögur: Orðstöðulykill og texti* [The Sagas of the Icelanders: Text and Concordance], (Reykjavík: Mál og menning, 1996).

6. For a recent and thorough review see Seán McGrail, *Ancient boats in N.W. Europe: The archaeology of water transport to AD 1500* (London and New York: Longman, 1987).

7. Anne Stine Ingstad, *The Discovery of a Norse Settlement in America: Excavations at L'Anse aux Meadows, Newfoundland 1961-1968* (Oslo, Bergen and Tromsø: Universitetsforlaget, 1977).

8. Vilhjálmsson, "Time-reckoning".

9. Cf. the first and second categories of sources in section 1.2 on sources.

10. A. S. Ingstad, *The Discovery;* Helge Ingstad, *The Norse Discovery of America, Vol. 2: The Historical Background and the Evidence of the Norse Settlement Discovered in Newfoundland* (Oslo, Bergen, Stavanger, Tromsø: Norwegian University Press).

11. Helle Degnbol et al. (eds.), *A Dictionary of Old Norse Prose, Vol. 1: a-bam* (Copenhagen: The Arnamagnean Commission, 1995), 733.

12. The noncomposite words ótta, nón, etc., do not have any clear-cut meaning in terms of other words or concepts.

13. Einar îlafur Sveinsson and Matthías Þórðarson (eds.), *Íslenzk fornrit IV: Eyrbyggja saga, Grænlendinga sögur*, (Reykjavík: Hið íslenzka fornritafélag,

1935), 246; Bragi Halldórsson et al., *Íslendinga sögur og þættir: Síðara bindi [II]* (Reykjavík: Svart á hvítu, 1986), 1097-8. Spelling is normalised as in the latter edition.

14. Magnus Magnusson and Hermann Pálsson (eds. and trans.), *The Vinland Sagas: The Norse Discovery of America*, (Harmondsworth: Penguin, 1965), 52-3.

15. Vilhjálmsson, "Time-reckoning", 72-4.

16. Holm-Olsen, *Konungs skuggsiá*, 10, here with standardized spelling.

17. My translation. The translation of Larson, *King's Mirror*, 94, goes as follows:

> Merchants are, however, scarcely able to note these changes, as the course is too swift; for the moon takes such long strides both in waxing and in waning that men, on that account, find it difficult to determine the divisions of its course. The sun, on the other hand, completes its course more slowly both in ascending and declining, so that one may mark all the stages of its course.

18. Sveinsson and Þórðarson, *Íslenzk fornrit IV*, 251; Halldórsson et al., *Íslendingasögur og þættir II,* 1100.

19. Magnusson and Pálsson, *Vinland Sagas*, 56, have the following for the last sentence: "On the shortest day of the year the sun was already up by 9 a.m., and did not set until after 3 p.m." Although the meaning is strictly speaking not the same it may be so in practice because of the uncertainty in observation and interpretation.

20. In other words 45 degrees east and west of south. Some scholars tend to read this as 67.5 degrees each way which would bring the maximum latitude much further south. However, the arguments for this reading as such are not convincing and for those interested in finding a maximum latitude, the logic of preferring that reading does not apply.

21. In the northern hemisphere of the earth the sun does not rise at winter solstice further than 66.5 degrees from south. Hence it is not obvious that people would describe this in different words even if they were in the tropics. In other words, the statement yields no southern limit for the location.

22. Søren Thirslund, "A presumed sun compass from Narsarsuaq", in Christen Leif Vebæk, *The Church Topography of the Eastern Settlement and the Excavation of the Benedictine Convent at Narsarsuaq in the Uunartoq Fjord* (Meddelelser om Grønland, Man and Society 14) (Copenhagen: The Commission for Scientific Research in Greenland, 1991). Cf. also Uwe Schnall, *Navigation der Wikinger: Nautische Probleme der Wikingerzeit im Spiegel der schriftlichen Quellen*, (Schriften des Deutschen Schiffahrtsmuseums, Band 6), (Oldenburg/ Hamburg: Gerhard Stalling, 1975), 85-8.

23. Thorkild Ramskou, *Solstenen: Primitiv navigation i Norden før kompasset*, (Copenhagen: Rhodos, 1969); Schnall, *Navigation*, 92-115; Allan A. Mills, "Polarization of Light from the Sky, and its Application to Timetelling and Navigation", *Bulletin of the Scientific Instrument Society*, No. 33, (1992), 8-14.

24. Oscar Albert Johnsen and Jón Helgason (eds.), *Saga Óláfs konungs hins helga: Den store Saga om Olav den Hellige I-II*, (Oslo: Kjeldeskriftfondet, 1941), 662; Bergljót S. Kristjánsdóttir et al., *Heimskringla III: Lykilbók*, (Reykjavík: Mál og menning, 1991), 19.

25. My translation.

26. Johnsen and Helgason, *Saga Óláfs*, 670-1; Kristjánsdóttir et al., *Heimskringla III*, 23.

27. My translation.

28. Mills, "Polarization", 12, is similarly nonplussed.

29. In a report on ostensibly 9th century events in the manuscript *Hauksbók* from around 1300 it is explicitly stated that "at that time no seamen had a lodestone in the Nordic lands"; Jakob Benediktsson (ed.), *Íslenzk fornrit I: Íslendingabók, Landnámabók* (Reykjavík: Hið íslenzka fornritafélag, 1986), 37. This comment is not found in *Sturlubók*, another manuscript of the *Book of Settlements* which is thought to be some 50 years older.

30. Janet Bately (ed.), *The Old English Orosius*, (London, New York, Toronto: Oxford University Press, 1980), 13-14.

31. B. Thorpe (ed.), "Alfred's Anglo-Saxon Version of Orosius with a Literal English Translation and an Anglo-Saxon Alphabet and Glossary", Reinhold Pauli (trans.), in *The Life of Alfred the Great*, (London: Henry G. Bohn, 1853), 249.

32. See McGrail, *Ancient Boats*, 263-4 and references therein.

33. Previously often called "family sagas" outside Iceland, cf. Jónas Kristjánsson, *Eddas and Sagas*, 203 ff.; 207 for this particular point.

34. Geoffrey Jules Marcus, *The Conquest of the North Atlantic*, (Woodbridge, Eng.: Boydell Press, 1980), 108-9, 209.

35. Benediktsson, *Íslenzk fornrit I*, 33. - Spelling is normalised in the same way as in other quotations in the paper.

36. "straight west to Greenland" in another manuscript, which fits better in with most of the other data.

37. "five" in another manuscript, which is more plausible.

38. Jan Mayen is much closer to Iceland than the present Svalbard. Thus you would expect that Jan Mayen was discovered before Svalbard. Since there is no other term for Jan Mayen in the Old Norse literature this seems to be the most natural identification.

39. My translation.

40. i.e. that "1 day" may mean between 0.5 and 1.5 days.

41. Magnusson and Pálsson, op.cit..

42. Sveinsson and Þórðarson, *Íslenzk fornrit IV*, 246-7; Halldórsson et al., *Íslendingasögur og þættir II*, 1097.

43. Magnusson and Pálsson, *Vinland Sagas*, 53-4

44. A. S. Ingstad, *The Discovery*; H. Ingstad, *The Norse Discovery*.

45. Birgitta Linderoth Wallace, "The L'Anse aux Meadows Site", Appendix VII in Gwyn Jones, *The Norse Atlantic Saga: Being the Norse Voyages of Discovery and Settlement to Iceland, Greenland, and North America*, (Oxford: Oxford University Press, 1986), 300-2; Gwyn Jones, *A History of the Vikings*

(Oxford: Oxford University Press, 2nd edn 1984, repr. 1990, 303-6; Magnusson and Pálsson, *Vinland Sagas*, 42; Jónas Kristjánsson, *Eddas and Sagas*, 15-16.

46. Holm-Olsen, *Konungs skuggsiá*, 10. An "ætt" here clearly means an azimuth angle of 45 degrees or one eighth of the full circle of the horizon (it is even conceivable that *átta* and *ætt/átt* are etymologically related, see e.g. Ásegir Blöndal Magnusson, Íslensk orðsifjabók [Icelandic Etymologial Dictionary, Reykjavík: Mál og menning, 1989]). The statement implies that the sun moves with equal speed relative to the horizon. As discussed in subsection 2.1, this is not a bad approximation when you are at or above 60 degrees of latitude.

47. My translation.

48. Beckman and Kålund, *Alfræði II*, 7.

49. My translation.

50. In the sagas of the Icelanders the term "misseri" occurs 89 times whereas there are only 10 occurrences of the term for "year" (Eiríkur Rögnvaldsson et al. (eds.), *Orðstöðulykill*.

51. Magnússon, *Orðsifjabók*, under "misseri".

52. See Vilhjálmsson, "Time-reckoning".

53. In some softer contexts, the meaning might soften such that the length of a single "dægur" might be unspecified whereas there would still be two "dægur" in each 24 hours. This corresponds to the fact that in the north the lengths of bright day and dark night are quite variable.

54. Kristian Kålund (ed.), *Alfræði íslenzk: Islandsk encyklopædisk litteratur, I. Cod. mbr. AM. 194, 8vo,* (Copenhagen: Samfund til udgivelse af gammel nordisk litteratur), 12-13.

55. literally "red"

56. My translation.

57. The actual difference in latitude is around 3 degrees. However, the direction is not at all straight NS so that the shortest distance as the crow flies is actually close to 4 degrees.

58. Beckman and Kålund, *Alfræði II*, 125.

59. My translation.

60. McGrail, *Ancient Boats*, 263-4.

61. Cf. the famous voyage of Þórarinn Nefjólfsson from the area around Staður to Eyrar in Iceland. He got an extremely favorable wind and was only 8 "dægur" on the way, corresponding to 6.5 "dægur" from Staður to Horn. He stated later on that the trip had only taken 4 nights. So, *in this context*, it is clear that "dægur" means 12 hours which, however, does not contradict our interpretation of "dægurs sigling" in general sailing itineraries. Cf. Snorri Sturluson, *Heimskringla I*, eds. Bergljót S. Kristjánsdóttir et al., (Reykjavík: Mál og menning, 1991), 405.

Ibn `Arabî's Cosmogony and the Sufi Concept of Space and TIme

by Samer Akkach

Since the time of the Prophet Muhammad in the 7th century until the mid-19th century various Muslim scholars had compiled and reported a wealth of traditions and mythical stories which dealt with the cosmogonic process and the structure of the universe. These traditions were authenticated by attribution to either the Prophet or one his immediate companions, an authentication which is contested, in various ways, by many Islamic schools. Regardless of their authenticity, however, these widely reported traditions represent a mythological apparatus of immense symbolic and social power. Many of these traditions and prophetic sayings appear, in one form or another, in most traditional chronicles, which normally begin their historical narratives from the time of the creation, in narratives, in religious and folk literature, and in many forms of social practices.[1] But cosmogony as a dimension of medieval Islamic metaphysics was dominated by the Sufis who developed it as an integral part of their mystical pursuit for the ultimate meanings of existence. The Sufis employed these traditions together with their Quranic[2] references to present a comprehensive and articulate view of the universe that encompasses every aspect of both the Divine and human levels of existence. The central figure responsible for developing Islamic cosmogony into its highest and most articulate state is Abû Bakr Muhammad b. `Alî b. Muhammad Ibn Al-`Arabî A1-Hâtimî Al-Tât'î, known as Muhyî al-Dîn Ibn Al-`Arabî, abridged as Ibn `Arabî. Born in Murcia, Spain, in 560/ 1162, Ibn `Arabî is arguably one of the most prolific and influential figures in the history of Islam. Among Sufis he is known as *al-shaykh al-albar* (the "Greatest Master").[3] In his writings and teachings, Ibn `Arabî presents an elaborate and cohesive view of the world based on a very intricate ontological structure.

His ontology brings together harmoniously a wealth of philosophical, theological, scientific, linguistic, metaphysical and mystical knowledge, developing it into a cohesive multi-dimensional whole with a unique level of intricacy and profundity. Islamic cosmogony after Ibn `Arabî revolved almost entirely in the sphere of his dominating thought and vision. This essay aims to explain, with reference mainly to *al-Futûhât al-Makiyyah*, the basic elements of Ibn `Arabî's cosmogony and his spatio-temporal structure of the world.[4] Special emphasis is placed on the Sufi concept of time and eternity, which is based on his teachings, with particular reference to a treatise on time and eternity by the eminent 14th century Sufi master `Abd Al-Razâq al-Qâshânî. Al-Qâshâni is a follower of Ibn `Arabî's school and a well known commentator of his work. The text of the treatise used is in manuscript form and is obtained from at Al-Zâhiriyyah library in Damascus. Other manuscripts from the same library are also used to present alternative diagrams to those available in the published version of Ibn `Arabî's *Futûhât* (see manuscripts cited list at end of article). This article uses interpretive exposition in order to present a cohesive view of a large body of knowledge based on an intricate web of complex Sufi ideas. This interpretive method is important for understanding of medieval cosmogony and views of the universal structure of space and time in general and the complex Sufi ideas in particular. This moment in Islamic metaphysics corresponds loosely with the late medieval period in Europe.

Patterns of Universal Manifestation

In Ibn `Arabî's cosmogonic schema, the creation of the world and the differentiation of primordial Unity are two expressions for one thing. Using geometrical and mathematical metaphors to illustrate his ideas, Ibn `Arabî assimilates the process of creation and differentiation to that of the proceeding of all geometrical figures from the undifferentiated point and all numbers from the unqualified zero. Human logic projects a certain order and limitation on this process. From nothingness there must have been a point first and then an extension before there could be a figure, and likewise from the zero there must have been a unity and then addition before there could be a number. The powerful logic of analogy was widely used in medieval Sufi literature to illustrate the process of universal manifestation. The world in its entirety was seen as an enormous sphere and was thus analogously compared with a geometrical circle within which the determination of geometric figures was compared with the indefinite differentiation of possible beings.[5] Geometrically, the circle is seen as the first comprehensible form of unity which the incomprehensible point takes on in the process of manifestation.

Metaphysically, the circle represents the first polarisation that differentiates the Unity of the Essence from the multiplicity of the Names, but without distinguishing the Names one from another, just as the points of the circumference are all alike and equally related to the principial point-centre. As the Divine Names become differentiated and distinguished from one another, they manifest an infinite number of patterns. These patterns crystallise the infinite variety of combinations the Divine Names generate. The Sufis compare these patterns to the indefinite number of geometrical configurations that can be inscribed within a circle, each, as it were, crystallising one of the Divine Patterns.

A square, for example, crystallises geometrically the pattern of quadrature, which in turn crystallises the relationship that binds any four distinct yet related Divine Names, such as the 'Living', the 'Knowing', the 'Willing' and the 'Powerful'. For Sufis, combinations of the Divine Names constitute the regulating pattern of existence, varying according to the subjects they designate (see figure 1, page 138). The creation of the world, for example, by virtue of the nature of the creative act, requires a pattern different from that required for the subsistence of the world after it has been created. Likewise, different modalities of the Divine reveal different patterns. A different combination of Attributes is required to know God as the Creator of the world from the ones required to know Him as the Self-sufficient. The Sufis discern a structure in these infinite varieties of patterns, based on hierarchical order in both single and combined Divine Names. 'Quadrature' and 'Triplicity' occupy, with regard to the creative process, a primary position. Ibn `Arabî considers the principles of triplicity and quadrature to be the two primary means of ordering, which project at different levels of manifestation in different modalities.

In the corporeal world, the lower end of the hierarchy, triplicity and quadrature crystallise in the manifestation of space and time. Space in its three dimensions -- length, breadth and depth -- is seen as a crystallisation of the sacred ternary, whereas time in its fourfold division -- the four seasons and the four temporal measures, day, week, month and year -- is seen as a crystallisation of the sacred quaternary. The coexistence and concurrence of space and time is seen as a phenomenal expression of the binding re-lationship between these two sacred patterns of universal manifestation.

Quadrature: The Pattern of Proliferation

According to Ibn `Arabî, creation requires the following combination of Divine Names: the 'Living' (*al-hayy*), the 'Knowing' (*al-`alîm*), the 'Willing' (*al-murîd*) and the 'Powerful' (*al-qadîr*). These manifest the Attributes of Life, Knowledge, Will and Power, the four principal and indispensable qualities required for any creative act.[6] This combination with its inherent order constitutes the sacred pattern of quadrature, after which all quadratures in the world are fashioned.[7] Ibn `Arabî explains the order inherent in this quadrature: Life, he says, is the primary attribute, being the very condition of existence, while the other three are secondary. Life and Knowledge are also of a higher order than Will and Power. Life and Knowledge are therefore active in relation to Will and Power, which are passive. Ibn `Arabî says that since Life is the condition of Knowledge and Will is the condition of Power, Life corresponds to Will and Knowledge corresponds to Power, for Power is to Will as Knowledge is to Life.[8] The limits of the realms to which each of these Attributes extends in the creative process is the basis of this order. Life is the highest in the hierarchy because it is the very condition of all Attributes.[9] Knowledge follows, being capable of dealing with the realm of the Necessary Being and the realm of possibility, which includes that which can exist (*al-mumkin*) and that which cannot (*al-mahâl*). Then follows Will, being more restricted than knowledge in that it can only deal with the realm of the possible by giving preference for a possible being to exist or not to exist. Power is the last in the hierarchy, being the most restricted, since its function is to bring the possible being, which has already been given preference for existence, into existence.[10]

The differentiation of the first Divine Attributes of Life, Knowledge, Will and Power represent the frst stage in the cosmogonic process. The second stage begins with the manifestation of the Universal Intellect. With reference to a Prophetic tradition which says that creation started with the 'Intellect' (*al-`aql*) and followed by the creation of the 'Soul' (*al-nafs*) from the Intellect, Ibn `Arabî explains that God brought the First Intellect into existence from the attribute of Life and the Soul from the Attribute of Knowledge.

Accordingly, the Intellect is the Soul's condition of existence just as Life is Knowledge's condition of existence. The two passive realities related to the Intellect and the Soul are Nature and Matter.[11] 'Nature' (*al-tabî'ah*) occupies the state between 'Soul' and 'Matter'.[12] For Sufis, Nature, just like time, is something intelligible that can be known yet has no essence (*`ayn*): an intelligible force that comprises the four realities of Heat *harârah*), Cold (*burûdah*), Dryness (*yubûsah*) and Moistness (*rutûbah*), whose effects, and not themselves, are sensed in the created, natural bodies.[13] Heat and Cold are active principles while Dryness and Moistness are passive. Dryness is from Heat while Moistness is from Cold. Heat is from the Intellect, and the Intellect from Life, hence the symptom of life in the sensible body is heat. Cold is from the Soul, and the Soul from Knowledge, hence, knowledge, when settled, is usually described in Arabic as the 'cold of certainty' and as 'snow'.[14] As Dryness and Moistness are passive with regard to Heat and Cold, Will demands

Dryness because it belongs to its state, while Power demands Moistness. Ibn `Arabî assigns to 'Power' the role of bringing into existence, and hence he associates Power with manifesting the qualities of heat and moistness, seen as the primary symptoms of life, in all living things.[15]

The third stage in the cosmogonic process is the creation of the four ideal elements: Fire, Air, Water and Earth. Ibn `Arabî explains the manifestation of this stage with reference to the Quranic verse, "...the heavens and the earth were of one piece [*martûqah*], then We parted them, and We made every living thing of water?" (XXI, 30). He says that heaven and earth were first manifested in the form of an undifferentiated, indistinctive mass (*martûqah* lit., 'stitched together'), then God 'unstitched' the sewn mass in order to distinguish them one from the other, and Water was the tool.[16] Fire, Air and Earth followed. Water is the primary element, just as Life, to which it relates, is the primary Divine Attribute. To create these four ideal Elements, Ibn `Arabî explains, God combined Heat and Dryness and there was the simple and ideal Fire; Cold and Dryness and there was the simple and ideal Earth; Heat and Moistness and there was the simple and ideal Air; and Cold and Moistness and there was the simple and ideal Water. (See figure 2, page 139).

Triplicity: The Pattern of Formation

In his cosmogonic process Ibn `Arabî distinguishes two modes of creation: creation by imagining, knowing or conceiving, that is, by designation or determination (*taqdîr*), and creation by producing and manifesting, that is, by bringing into existence or formation (*takwîn*). The former is an eternal, creative act that does not involve physical production; while the latter is a creative act that does involve physical production. The former co-exists with God's knowing of the non-existent world; whereas the latter co-exists with the bringing of the world from non-existence into existence, from potentiality into actuality.[18] The distinction between these two creative modalities is evident at the linguistic level where two distinct Arabic terms for 'world', *`âlam* and *kawn*, are available. The former derives from the verb *alima*, 'to know', and the noun *`ilm*, 'knowledge', indicating that the phenomenon of the 'world' (*al-`alam*) was originally perceived as being directly related to the notion of knowledge.[19] The latter means literally *hadath* (from which the terms *mahdath* and *hudûth* designate 'ephemeral existence' and 'temporality' as opposed to *qadîm* and *qidam*, which designate 'Divine Primordiality' and 'Eternity'), that is, 'something new', 'a novelty', 'an unprecedented thing', 'occurrence', and from which the verb *kawwana* means 'to create', 'to form' and 'to bring into existence'.[20] Although Knowing is a creative act, it is not necessarily a physically productive one. Things may be created in the imagination without being produced in tangible forms. *Takwîn*, however, is necessarily a physically productive act, causing the designated thing to exist. Thus creation has two complementary modes: 'designative' through knowledge and imagination (*taqdîr*), and 'productive' through bringing into existence and formation (*takwîn*).

Whereas quadrature is seen as the pattern of proliferation and deployment and as the sacred pattern of creation with regard to *taqdîr*, triplicity is seen as the pattern of synthesis and formation and as the sacred pattern of creation with regard to *takwîn*. Thus triplicity complements the pattern of quadrature in crystallising the productive, as opposed to the designative, aspect of creation. In the Divine's productive act triplicity is manifested through the three principles of production (*takwîn*): the source, that is, the Divine Essence, the intention, that is, the Divine Will, and the instument, that is, the Divine Word 'Be' (the imperative 'Be', *kun* in Arabic, derives from the above mentioned noun *kawn*, 'world' or 'cosmos'). This relationship derives from the Quranic verse: "And our word unto a thing, when we intend it, is only that we say unto it: Be! and it is" (XVI, 40). This active triplicity is complemented by a passive triplicity in the created thing: its *thing-ness (shay'iyyah)*, its hearing (*samâ`*), and its obeying (imtithâl) the command of the creator concerning its creation. The creaturely triad corresponds with the Divine triad causing this thing to exist.[21] lbn `Arabî represents this triplicity by a circle where the centre corresponds with the Essence, the radius with the Will and the points of the circumference with the command 'Be'. The circle is thus seen as delineating the domain of all possible beings.[22]

The association of triplicity with the spatial condition of existence is seen in relation to another set of Divine Names. Ibn `Arabî distinguishes two complementary modalities whereby God may be known: the creative modality, in which God is attached to the created object, the world; and the modality in which God is totally independent and Self-sufficient. The former is the modality known through the Divine Names the 'Living', the 'Knowing', the 'Willing' and the 'Powerful'; whereas the latter, in which "Allah is Independent of all creatures" (III, 97), can be known through the Divine Names the 'Living', the 'Speaking', the 'Hearing' and the 'Seeing'.[23] Together, these two modalities constitute the following seven Divine Names: the 'Living', the 'Speaking', the 'Knowing', the 'Hearing', the 'Seeing', the 'Willing' and the 'Powerful'. These Names are unanimously accepted by all traditional Islamic schools as being the seven principal Divine Names, from which all other Names derive. They are known in traditional Islamic literature as the 'Mothers of the Names' (*ummâhat al-asmâ*). Here triplicity corresponds to the three dimensions of length, width and depth in two ways. First, through the correlation beween the three dimensions and number eight viewed as the first cubic number, the seven principal Names together with the Essence are seen as the Divine pattern of spatiality. Any spatial manifestation involves at least eight points, because a line requires at least two points, a plane requires at least two lines, and a three dimensional volume requires at least two planes, hence, eight is considered to be the first cubic number. The body acquires the quality of length from the line, the quality of width from the plane, and the quality of depth from the volume.[24] Ibn `Arabî says just as spatial manifestations require a minimum of eight points, the manifestation of the Divine Presence occurred with the Essence and the seven principal Attributes. Second, the seven principal Divine Names is also seen to correspond with the six spatial directions -- up and down, left and right, front and back -- which confine all spatial manifestations. Ibn `Arabî says that of these seven Names only six relate to the possible being, while the seventh, the 'Living', does not. The Name 'Living' is the condition for the existence of the other six Names and the Principle where they meet. So the six Divine Names -- the 'Speaking', the 'Knowing', the 'Hearing', the 'Seeing', the 'Willing' and the 'Powerful' -- determine the six directions, while the 'Living Essence' is the centre where all coincide.[25]

The Spatio-Temporal Structure of the World

Moving down the hierarchy from the level of Divine Realities to the level of existential realities, Ibn `Arabî elaborates an intricate structure through which the above patterns of manifestation find immediate expression. At this level Ibn `Arabî distinguishes two worlds: 'World of Command' (*`âlam al-amr*), that is, the supra-natural world, and the 'World of Creation' (*`âlam al-khalq)*, the natural world.[26] The 'World of Command', also referred to as the 'World of Spirits' (*`âlam al-arwâh*), is the highest level in the hierarchy of cosmic manifestation that comprises the simple, cosmic models, in the likeness of which things are fashioned. Ibn `Arabî illustrates the 'World of Command' as shown in figure 3 (page 140).

The encompassing circle represents the form of the primordial 'Cloud' *(al-`amâ')*, which contains the thirty-two 'Stations of the Angels Ecstatic with Love' (*maqâmât almalâ'ikah al-mahayyamah*), who, according to a tradition, "have been enraptured by the Majesty and Beauty of God ever since their creation."[27] They are described in the tradition:

> God has a white earth in which the sun takes thirty days to cross the sky, and each of these days is thirty times longer than the days of the lower world. That earth is filled with creatures who do not know that God has been disobeyed in the earth or that He has created Adam and Iblis.[28]

The 'Angels Ecstatic with Love' include the 'First Intellect' (*al-`aql al-awwal*) and the 'Universal Soul' (*al-nafs al-kulliyyah*). However, the Angels are distinguished from the Intellect and the Soul, which, though enraptured by God's Majesty and Beauty, are engaged in other responsibilities with regard to the created world.[29] The First Intellect is identified as the supernal 'Pen' (*al-qalam*), and the 'Universal Soul' (*al-nafs alkulliyyah*) as the 'Guarded Tablet' (*al-lawh al mahfûz*). Hence, the World of Command is described as the 'Exemplar of the World' (*nuskhat al-`âlam*) which contains the 'Luminous Inscriptions' of the Pen upon the Tablet. Within the Guarded Tablet are manifested the Soul's 'two faculties' (*al-quwwatayn*), the theoretical and the practical.[30] Next to the Guarded Tablet appears the state of 'Nature' (*martabat al tabî`ah*), which includes four principles arranged in two antinomical pairs: 'Heat-Cold' (*harârah-burûdah*) and 'Dryness-Moistness' (*yubûsah-rutûbah*). The 'Universal Matter' (*al-hayûlâ al-kull*) is shown in the form of a circle analogous to the encompassing circle of the Cloud.[31] Within the Universal Matter God revealed the natural world, the 'World of Creation' (*`âlam al-khalq*), which includes the Divine Throne, the Divine Footstool, the celestial Gardens and the corporeal world of heaven and earth. Universal Matter is the substance supporting the manifestation of the World of Creation being, in the manner of the *materia secunda*, determined in accordance with the conditions of our physical world.[32] In itself, Universal Matter, like Nature, has no essence and is recognisable only through the manifestation of forms in it (*ma`qûl ghayr mawjûd al-wujûd al-`aynî*), in the same way that Nature can only be recognised through the manifestation of its effects in natural objects.[33] figure 4 (page 141) shows the level of the Divine Throne.

Ibn `Arabî' says that the interaction between the Intellect and the Soul produced a 'child', whose father was the Intellect and whose mother was the Soul. God hovered over it with his Name the 'Compassionate' (*al-rahmân*) because parents never regard their child except with compassion.[34] The notion of the Divine Throne (*al-'arsh*) as the all-encompassing limit of the world derives from the Quran: "The All-compassionate sat Himself upon the Throne."(XX, 5) According to Ibn `Arabî, the Throne marks the threshold between the domains of simple and of composed cosmic entities, between the World of Command (*`âlam al-amr*) and the World of Creation (*`âlam al-khalq*). The Throne is considered to be a composed form because its production involves four Realities, namely, Nature, Universal Matter, Universal Body and Circularity. First, God caused Universal Matter to exist, and by virtue of it receiving length, breadth and depth it became the Universal Body. Nature then manifested its conditions (*hutm*) in this Body. Circularity was the first form this Body received, so there was the sphere (*al-falak*). Spatially, the Throne is described as a seat with four *arkân* (lit., corners) and four faces. The four *arkân* are its main pillars (*qawâ'im asliyyah*); however, in each of the Throne's four faces, there are many pillars that are equally distributed. God made the Throne hollow, encompassing all that is contained in the natural world.[35] Within the Throne there is then manifested the Footstool (*al-kursî*), upon which the Divine Feet rest.

The Throne is associated with two notions: that of Light (*nûr*) and that of Spirit (*rûh*). A tradition says that God created the Throne from His Light. So the Divine Light forms, as it were, the substance of the Throne. The Spirit (*al-rûh*), "...most often figured as a 'centre', a 'ray', a 'descent', a 'presence' or 'immanence'," resides at the centre of the Throne.[36] The Throne is the 'Sphere of spheres', the outermost, all-encompassing limit of our world; it is the circumference (*muhît*, lit., 'encompassing') of the body of the world, expressing the principles of 'totality' and of 'integration'; it is the circumference of which the Spirit (*al-rûh*) is the 'centre' and the Light (*al-nûr*) is the 'matter'.[37] Thus the Throne embodies the principle of circularity, reflecting the pattern of the Divine Presence, whose Essence corresponds to the central Spirit and whose Names correspond to the encompassing circum-ference.[38] The forms of the Throne and the Footstool crystallise the pattern of quadrature. The four bearers of the Throne correspond to the four 'Divine Pillars' -- Life, Knowledge, Will, and Power that support the Divine Presence. Many traditions say that God created the Throne from green emerald and its four pillars from red ruby.[39] The bearers of the Throne are four angels whose feet are in the nethermost earth and whose heads are in the Throne's domain. The four bearers have the forms of a man, a bull, an eagle and a lion.[40] The celestial rivers are laid out about the Throne in the pattern of quadrature. A tradition says that about the Throne there are four rivers: a river of sparkling light, a river of blazing fire, a river of shining white snow and a river of water.[41]

Ibn `Arabî says that inside the Footstool God created a 'circular, transparent body', which He divided into twelve sections. He called these sections 'towers' (*burûj*, i.e., 'mansions of the stars') as mentioned in the Quranic verse: "By the heaven, holding mansions of the stars."(LXXXV, 1)[42] This 'circular, transparent body' is called the *atlas* sphere, and is commonly referred to as the sphere without stars and the sphere of the constellations. Within this sphere God then created the 'sphere of the fixed stars' (*falak al-kawâkib al-thâbitah*). According to the 16th/17th century astrologer Baha'al-Dîn al-`Amilî, in his treatise *Tashrîh al-Aflâk* ('The Anatomy of the Spheres'), the *atlas* sphere and the sphere of the fixed stars are none other than the Throne and the Footstool respectively, the former being, as it were, their technical names, whereas the latter arc their Quranic equivalents.[43] In astrology, the sphere of the fixed stars is the cosmic container of the corporeal world of the heavens and the earth. This corresponds with the Quranic description: "His Footstool (*kursî*) includeth the heavens and the earth" (II, 255). On many occasions, Ibn `Arabî also identifies the *atlas* sphere with the Throne. Ibn `Arabî says that between concave surface of the *atlas* sphere and the convex surface of the sphere of the fixed stars, God positioned the celestial Gardens as shown in figure 5 (page 141). Ibn `Arabî distinguishes eight Gardens, seven of which are hierarchically ordered in seven levels with an eighth, the *al-wasîlah*, as the superior one in which all the other levels participate.[44] This hierarchical structure reflects the pattern of the Divine Presence: the seven Gardens correspond to the Seven Principal Names and the superior one, *al-wasîlah*, corresponds to the Essence.[45]

The *atlas* sphere has a duodenary structure which derives from multiplying the quaternary of Universal Nature by the ternary structure of the natural world.[46] Nature conditions the characteristics of the duodenary structure of the *atlas* sphere through its four ideal Elements, Fire, Air, Water and Earth, revealing the Igneous, Terrestrial, Aerial and the Aqueous signs. The Igneous is composed of Heat and Dryness, the Terrestrial of Cold and Dryness, the Aerial of Heat and Moistness, and the Aqueous from Cold and Moistness. As a result of this multiplication, all divisions of the *atlas* sphere which are related by the trigon have the same elementary nature, whereas all divisions that are related by the square have different elementary nature (see figure 6, page 142).

According to the Quranic description of the Gardens in the *sûrah* of 'The Constellations': "Lo! those who believe and do good works, theirs will be Gardens underneath which rivers flow" (LXXXV, 11), the pattern of the sphere of the constellations is seen to correspond to the inherent structure of the Gardens. Ibn `Arabî says that God situated the Gardens on an outer surface of the sphere of the fixed stars. The surface of this sphere is made of musk which forms the soil of the Garden. God then divided the Gardens into three divisions according to the three aspects each sign has: the primary Gardens are 'Gardens of Distinctiveness' (*jannât al-ikhtisâs*), the secondary Gardens are 'Gardens of Inheritance' (*jannât al-mîrâth*), and the tertiary Gardens are 'Gardens of Deeds' (*jannât al-a`mâl*).[47] Then in each division God made four rivers to flow, which when multiplied by the three divisions results in twelve rivers.[48] These four rivers symbolise the four sources of esoteric knowledge.[49] They are: a river of 'Unchanging Water' (*mâ' ghayr âsin*) symbolising the 'science of life' (*'ilm al-hayât*); a river of 'Wine' (*khamr*) symbolising the 'science of the spiritual states' (*'ilm al-ahwâl*); a river of 'Honey' (*'asal*) symbolising the 'science of divine revelation' (*'ilm al-wahî*) with its many kinds; a river of 'Milk' (*laban*) symbolising the 'science of the secrets' (*'ilm al-asrâr),* the kernel of all sciences, which God directly reveals to those who devote themselves utterly to Him.[50]

Divine and Human Duration

Sufis view spatial extension and temporal duration to be the fundamental conditions of all manifestations within the Divine Throne, and consider motion to be the generative principle of space and time. The coexistence and concurrence of space and time is the direct result of motion (*harakah*). The primordial motion was seen as a form or condition imposed by the Universal Soul on the Universal Body after admitting the form of the three dimensions.[51] Thus viewed, time was associated with the principial motion of the Universal Body. The Universal Soul, which selected the spherical form for the Universal Body because it was the noblest of all forms, chose for it the circular motion for the same reason.[52] Explaining why the principial motion had to be circular, Ibn `Arabî says that it was necessary for the Universal Body to move so that the principles of duration and extension can be manifested; however, since there was no vacuum except that which had already been filled by the Body, there was no space towards which the Body could move, so the Universal Body moved in its place about its centre.[53] The Sufis use the metaphor of a millstone to describe the motion of the Universal Body. While every part of a millstone moves from its place into another by the movement of the whole, the whole itself does not change its location by the movement of the parts. The circular movement was described by Ibn `Arabî as moving-still: still in regard to the whole and moving in regard to the parts.[54]

Ibn `Arabî envisions that the *atlas* sphere is the cosmic progenitor of space and time. The Arabic term *atlas* means literally 'effaced' or 'obliterated', denoting the idea of perfect homogeneity without distinguishable features. For this reason it is also referred to as 'the sphere without stars'. Motion coincides with the actualisation of this sphere, which, seen independently of the planetary heaven it contains, engenders a pure and undifferentiated spatio-temporal modality.[55] In its cosmic context, the *atlas* is encompassed by the sphere of the Footstool, and it encompasses the sphere of the fixed stars or of the moon's mansions, which englobes the corporeal domain. Being perfectly homogeneous and without planets, the *atlas* sphere in itself lacks any distinguishable reference point that could differentiate its incessant continuous movement into recurrent cycles. In order for its movement to be differentiated, it must therefore be related to other cosmic elements either from within or from without this sphere. Ibn `Arabî differentiates the movement of the *atlas* sphere by way of relating it, from within and from without, to two distinct referents. From within, he relates the movement of the *atlas* sphere to the sphere of the fixed stars and the planets; and from without, to the position of the Divine Feet on the Footstool. By means of these two referents, Ibn `Arabî distinguishes two different modes of time: Divine and Human.[56] The Divine mode is that which is determined from without the *atlas* sphere by the Divine Feet, and as such it is knowable only to God. The Human mode is that which is determined from within the *atlas* sphere by the movements of the stars, and as such it is knowable to man.[57] The Divine mode forms an archetypal model according to which the human mode is articulated.

First of all, Ibn `Arabî distinguishes the movement of the stars from that of the *atlas* sphere. He says that the revolution of the stars within the *atlas* sphere is subject to two kinds of movements: natural (*tabî`iyyah*) and forcible (*qasriyyah*).[58] Natural movement is that whereby the stars traverse the *atlas* sphere from West to East, as seen by the eye. Forcible movement is the reverse one whereby the stars move with the movement of the *atlas* sphere from East to West.[59] It is like an ant on a rug, he says, which is being pulled west-wards as the ant is moving eastwards: the ant is simultaneously moving to the East and to the West. Likewise the stars, at the same instant in which they are naturally moving from West to East they are also forcibly moving from East to West (see figure 7, page 142).

By means of the position of His Feet on the Footstool, God differentiates the primordial motion of the *atlas* sphere into seven recurrent cycles, or 'Days', in accordance with the seven principal Names or Attributes already discussed, which constitute the integral pattern of the Divine Presence. Ibn `Arabî explains:

> By the existence of the *atlas* sphere there occurred the seven days, the months and the years. But these times were not determined until after God created inside this sphere the signs whereby these times were distinguished. The only duration this sphere determines is the 'Day': it is one cycle determined by the position of the Foot on the Footstool. So it is determined from above, and the measure of one complete cycle is called *yawm* ('Day'). Because of the homogeneity of

> this sphere, this 'Day' is known only to God (-most transcendent). The starting point of its movement coincides with the first degree of Gemini, which is among the aerial signs, facing the Foot. The first 'Day' manifested in the world was in the first degree of Gemini, and that 'Day' was called *al-ahad* (Sunday).[60]... Then upon the end of the first cycle the sphere started another movement... This second movement was called *yawm al-ithnayn* (Monday). And so [the cycles proceeded] until the seven cyclical movements were completed, with one Divine Attribute determining each movement. As the [principal] Attributes are no more than seven, the days of the *al-dahr* (aeon) cannot be more than seven, not even by a day, for there is nothing that may necessitate it. Thus, the ruling returned to the first Attribute that rotated the sphere [again] and the name *ahad* became associated with it... [For the new cycle], however, it was more appropriate to be the eighth with regard to the cycles overall, but since its existence was caused by the first Attribute itself its name did not change. Similarly, the cycle which followed, and so was the one which followed until the seven cycles [were once again completed].[61]

According to Ibn `Arabî, each Divine Name or Attribute engendered one entire cycle or Day, during which period this Attribute dominated over the other Attributes, causing its characteristic to inhere in all creatures. This is to say that during the cycle which was engendered, and hence, dominated by the Divine Attribute of 'Hearing', for example, all creatures received the characteristic of hearing by virtue of which they become able to hear, and the same for the rest of the principal Attributes. As the creation coincided with the utterance of the Primordial Word, hearing was the first characteristic the creatures received so that they became able to respond to the Divine Command 'Be!' The creative process took place in the following order:[62]

Hearing	Sunday	= *al-Ahad*
Living	Monday	= *al-Ithnayn*
Seeing	Tuesday	= *al-Thulâthâ'*
Will	Wednesday	= *al-Arbi`â'*
Power	Thursday	= *al-Khamîs*
Knowledge	Friday	= *al-Jumu`ah*
Speech	Saturday	= *al-Sabt*

Representing the governing patterns of universal manifestation, these seven Divine Attributes, on the one hand, determine the cyclical revolution of time engendered by the primordial motion of the *atlas* sphere, and on the other, necessitate the creation of seven planets, whereby the already determined motion from without by God may then be further differentiated from within by man. The aim of this complex schema is primarily to resolve the paradox of the 'six days of the creation'. The Quran says: "Lo! your Lord is Allah Who created the heavens and the earth in six Days."(VII, 54). If the existence of time is itself based on the movement of the heavenly stars or the existence of the heavens and earth, how could the heavens and earth be created in six days? If the 'day' mentioned in this verse is that temporal duration which we experience between two successive risings of the sun, no days should have existed before the existence of the stars, and the six days of the creation remain inexplicable in terms of time. By means of his schema of the two modes of time, Ibn `Arabî' was able to maintain that space and time already existed before the creation of heaven and earth. Not only that, but a 'Day' as a measuring unit was also differentiated in time by reference to the position of the Divine Feet on the Footstool. This 'Day' is a Divine Day -- "a Day with Allah is as a thousand years of what ye reckon" (XXII, 47) -- and not a human day, which nonetheless reflects its Divine model. Although corporeal time reflects the Divine model, this in no way suggests that the duration and the nature of corporeal time is the same as Divine time. In Divine time there are neither day-times, nor night-times, nor weeks, nor months, nor years, nor seasons, for all of these relate to the sun and the moon.[63] There is, instead, pure duration of only one ever-recurring Day.

Time and Eternity

In a short treatise on time and eternity, `Abd Al-Razâq al-Qâshânî, the eminent 14th century Sufi master and well known commentator on Ibn `Arabî's work, articulates Ibn 'Arabî's teachings on cosmogony and the manifestation of space and time in a lucid and structured manner.[64] Focusing on time and eternity, Al-Qâshânî discusses the Divine and human order of time, showing the measures of the 'Divine Days' (*al-ayyâm alilâhiyyah*) and the duration of the 'Eternal Year' (*al-sanah al-sarmadiyyah*). His primary Quranic references are: "And they demand of thee to hasten the chastisement! God will not fail His promise; and surely a day with they Lord is as a thousand years of your counting." (XXII, 47); "He directs the affair from heaven to earth, then it goes up to Him in one day, whose measure is a thousand year of your counting." (XXXII, 5); and "The angels and the Spirit ascend unto Him in a Day whereof the span is fifty thousand years." (LXX, 4)

Al-Qâshânî constructs a hierarchical view of time wherein he distinguishes three types of duration: eternal (*sarmadî*), aeonian (*dahrî*) and temporal (*zamanî*). These types correspond in an inverse manner to the levels of differentiation of Divine realities in the hierarchy of existence. Using the metaphor of a pyramid, the process of differentiation may be understood as proceeding from the unity of the apex, the 'Essence' (*al-dhât*), to the infinite multiplicity of the base, whereas the associated durational articulations proceed from the infinite expansion of the base to the unity of the apex, the 'now' (*alân*). All begins with the Divine Essence (*al-dhât al-ilâhiyyah*) whose duration (*imtidâd*), or the period of Its subsistence, is indeterminable. In the state of transcendental isolation, or absoluteness, the Essence has no attribute, no name, no order, that is, no affirmative description whereby it may be known or determined. Determination occurs with the manifestation of the Names, whereby the Essence descends from Its state of 'Absoluteness' (*al-ahadiyyah*) to the state of the 'Unity' (*al-wâhidiyyah*). This descent reveals a duality, a relationship and an affirmative order, by virtue of which a beginning is conceived. The polarity of the two states enables us to conceive of the notion of absolute transcendence associated with the first, higher state, the notion of unity associated with the second, lower state, and the notion of beginning associated with the process of polarisation which leads to the distinction between the two states. The notion of beginning implies an end and a duration extending in between. Al-Qâshânî defines the duration associated with the first manifestation as *alsarmad* ('eternity'): it is the condition that defines the relationship between the Essence and the Names. Eternity extends from *al-âzâl* (pre-eternity) to *al-abad* (post-eternity) and qualifies the durational condition of the Divine Presence. The Divine Presence is the presence of the Divine Names, and hence Eternity defines the relationship each Name has with the Essence. In other words, Eternity qualifies God's presence in relation to the Names the Knower, the Willing, the Powerful, etc., but in relation to the Essence, God is beyond eternity. Al-Qâshânî uses the term *azaliyyat al-âzâl* ('pre-eternity of pre-eternities') to metaphorically describe the condition of the Essence which in reality is beyond every condition, even the condition of Absoluteness. He uses the term 'pre-eternity of pre-eternities' only to affirm the precedence of the state of 'Absoluteness' (*al-ahadiyyah*) over the state of 'Unity' (*al-wâhidiyyah*). The plurality in 'pre-eternities' relates to the Divine Names: "if we consider the 'pre-eternity of pre-eternities' to be the 'Absolute Primeness' (*al-awwaliyyah al-mutlaqah*) that has no multiplicity, then the Divine eternity can be seen as multiplied by the multiplicity of the Names." Accordingly, Al-Qâshânî durational articulations begin with the manifestation of 'Unity', the state which proceeds from Absoluteness and whose pre-eternal duration has a beginning, which is the 'pre-eternity of pre-eternities' that marks the beginning of the 'Eternal Year' (*al-sanah al-sarmadiyyah*).

Below the Presence of Divine Names (*al-hadrah al-ilahiyyah*), identified with the all-encompassing Name Allah, there manifests the 'Presence of Lordship' (*hadrat alrubûbiyyah*), identified with the Name *al-rabb* (the Lord). This Presence depends in its absolute control over the created world on the power of the Divine Presence, and hence, its pre-eternity proceeds from the pre-eternity of the former. Al-Qâshânî considers the manifestation of the Presence of Lordship to be 'delayed' from the manifestation of the Divine Presence exactly in the same way the latter is 'delayed' from the Presence of the Essence. The principial duration (i.e., the duration of the subsistence of the Essence in the state of Absoluteness), he says, extends from the 'pre-eternity of pre-eternities' to the 'post-eternity of post-eternities' as one, unqualified and undifferentiated continuum without either beginning or end or division or proportion. Upon the projection of the attributive determination, this duration becomes articulated into the durations of the Names, and in turn those of the Names into the 'durations of the Lordship' (*al-imtidâdât al-rubûbiyyah*), which he calls *al-dahr* ('aeon').

Al-Qâshânî further explains that at our temporal level the undifferentiated, unqualified continuum can be analogously compared with the duration of the first pure spherical revolution. To understand this, one should imagine a perfectly homogenous sphere revolving in its place around its centre. Without identifying a reference point on, inside or outside the sphere this revolution cannot be differentiated or determined. It is one continuous motion without a beginning or an end, it is what Al-Qâshânî calls the 'absolute time' (*al-zamân al-mutlaq*). But if one considers a point of reference a recurrent cycle can be identified and the incessant continuum can be differentiated into measurable durations. This is how our temporal measures are determined. The year is differentiated by the movement of the sun whereby it traverses all parts of the utmost sphere of constellation till it returns to the starting point. By this movement the first pure duration is articulated into years; the year, in regard to the sun's traversing the constellations, into months; the months into weeks; the weeks, in regard to the sun's return to the first point in its diurnal movement, into days; the days into hours; the hours into minutes; the minutes into seconds; then into thirds till 'now', which is to time what the geometrical point is to a line. Al-Qâshânî says that the 'now', which can be interpreted as the present time, is the shortest part of time that, being too short, cannot be further divided except in the imagination (*al-wahm*). And in that they are all recurrent cycles of measurable duration, any one of these temporal divisions can be called 'day'. The 'now' can therefore be considered as the shortest day, while the year as the longest, and in the same way that the shorter duration is normally used to measure out the longer, like when we use the hours to measure out the days, the days to measure out the months, the months to measure out the years and the years to measure out the absolute time, so likewise, the absolute time, being the shortest pre-eternal duration, is used to measure out the aeon (*al-dah*r), and the aeon is used to measure out eternity (*al-sarmad*).

The manifestation of the Presence of Lordship, Al-Qâshânî explains, coincides with the manifestation of the spatio-temporal condition, that is, the creation of our world. Accordingly, the temporal durations, or the 'days', of the present world, relate to the Presence of Lordship, which extend from the beginning till the end of time. Being beyond the confines of the measure of the first temporal motion, the 'days' of *al-dahr* are, Al-Qâshânî says, longer than the temporal durations. With reference to the Quranic verse: "...and surely a day with thy Lord is as a thousand years of your counting."(XXII, 47), he works out the 'aeonian durations': the 'day' to measure one thousand year (1,000), the 'week', being determined by the seven Primary Divine Names and the seven planetary heavens, to measure seven thousand years (7,000), the 'month' thirty thousand years (30,000), and the 'year' to measure three hundred and sixty thousand years (360,000). And with reference to the Quranic verse: "To Him the angels and the spirit mount up in a day whereof the measure is fifty thousand years."(LXX, 4) Al-Qâshânî works out in a complex way the measure of the 'eternal durations': the 'day' to be fifty thousand years (50,000), the 'week' three hundreds and fifty thousand years (350,000), the 'month' one million and five hundred thousand years (1,500,000), and the 'year' eighteen million years (18,000,000).

In conclusion, it is pertinent to mention that Ibn `Arabî's cosmogony and universal structure of space and time are an integral part of his doctrine of the 'Unity of Being' (*wadat al-wujûd*), through which he views the whole universe as so many refractions of One Reality. This One Reality proliferated to manifest the multiple States of Being, which are at once degrees of Reality and degrees of Unity. The One is the Principle whence multiplicity emanates, and the Divine pattern governing this process of emanation is the 'thread' (*nizâm*) which ties this multiplicity to its Principle. Multiplicity is revealed as so many levels of existence and so many layers that at once veil the supreme One and filters the intensity of Its dazzling light. As all the levels of existence originate from the One, each level bears the imprint of Unity. Harmony results from the reflection of the Primordial Unity in the multiplicity of the created world. The immediate cause of the manifestation of a particular level of existence always lies beyond the conditions of this manifested level. The cause of the manifestation of heaven and earth lies in the Footstool, that of the Footstool in the Throne, that of the Throne in Universal Matter, that of Universal Matter in Nature, that of Nature in the Universal Soul, and that of the Universal Soul in the Universal Intellect. So each level contains the principle of the one below it, and hence contains the realities of this inferior level in a non-differentiated mode. Accordingly, the cosmos is the Divine schema of inexhaustible correspondences. There is nothing in the universe that exists in isolation, all things are interrelated both horizontally, when considered at a particular level, and vertically, when considered at different levels. The multi-levelled cosmos is a whole made up of a multitude of superimposed layers of interrelated elements. Each level of existence reflects the pattern of the one that lies above it, while acting as a model for the one that lies below it, thus forming a continuous chain that has God at one end and man at the other. The Sufis climb up this chain in their contemplative ascension. Al-Qâshânî concludes his treatise by saying whoever rises to the 'Presence of Unity' (*al-hadrah al-wâhidiyyah*) steps out of the Days of Lordship into the Divine Days in the Eternal Year (*al-sanah al-sarmadiyyah*), and whoever steps out of eternity and reaches the 'Presence of Absolutness' (*al-hadrah al-ahadiyyah*) transcends all modal-ities of measurable time. He soars high above all states of being and his time become one. "God alone subsists after the annihilation of all the creatures, and that is the 'Truthful Day' (*al-yawm al-haqq*)."

University of Adelaide

Manuscripts Cited

The manuscripts listed below are from the Dâr al-Kutub al-Zâhiriyyah in Damascus, Syria Currently they are held in the national library, Maktabat al-Asad. The manuscripts are identified here by the numbers of tbe Zâhiriyyah library, according to which they are indexed. Full details of each manuscript are given in the published index referred to at the end of each item.

1. Al-Amidî, Ibrâhîm al-Qarmani (MS.6102) Risâlah fî Bayân `Adad al-Samâwât wa -l-Ardîn wa -l-`Arsh wa -l-Lawh wa -l-Qalam wa `Ajâ'ib al-Samâwât wa Makân al-Jannah wa -l-Nâr (32 folios) (cited in Khourî, Ibrâhîm, *Fihris Makhtûtat Dãr al-Kutub al-Zãhiriyyah*: *'Ilm al-Hay'ah wa Mulhaqâtihi*. Damascus, Matbû`at Majma` al-Lughah al-`Arabiyyah, 1969, 23).

2. Al-`Amilî, Baha' al-Dîn (MS.3103), *Tashrîh al-Aflâk*. 11 folios (cited in Khourî, *Fihris*... 1969, 16).

3. Ibn `Arabî;, Muhyî al-Dîn (MS.1328), Al-Futûhât al-Makiyyah fî Asrâr al-Malikiyyah wa al-Malakiyyah. (Vol. 1) (cited in Al-Malîh, Muhammad Riyâd, *Fihris Makhtûtât Dâr al-Kutub al-Zâhiriyyah: al-Tasawwuf*: Damascus, Matbû `at Majma` al-Lughah al-`Arabiyyah, 1978, 2: 368).

4. Ibn `Arabî, Muhyî al-Dîn (MS.1330) *Al-Futûhât al-Makiyyah fî Asrâr al-Malikiyyah wa alMalakiyyah*. (Vol. 3) (cited in Al-Malîh, *Fihris*...1978, 2: 368-70).

5. Ibn `Arabî, Muhyî al-Dîn (MS.3975) *Inshâ' al-Dawâ'ir wa -l-Jadâwil*. From folio 4 to 15 (cited in Al-Malîh, *Fihris*... 1978, 1:124-6).

6. Al-Qâshânî, `Abd al-Razzaq (MS.5258) Bayân Miqdâr al-Sanah al-Sarmadiyyah wa Ta`yîn al-Ayyâm al-Ilâhiyyah. From folio 90 to 93 (cited in Al-Malîh, *Fihris*... 1978, 1:213-4).

7. Al-Suyûtî, Jalâl al-Dîn (MS.5881) *Al-Hay'ah al-Saniyyah fî al-Hay'ah al-Sunniyyah*. 19 folios (cited in Khourî, *Fihris*... 1969, 37).

Notes

1. These traditions can be found in the history accounts of most traditional Muslim chroniclers, like al-Azraqî, al-Tabarî, Ibn al-Athîr, al-Mas`ûdî, Ibn Kathîr, etc, some of whom (e.g. al-Tabarî and Ibn al-Athîr) are also eminent religious scholars. It is interesting to observe how the reported myths play double roles as historical events and well as religious concepts.

2. For simplicity the transliterations 'Quran' and 'Quranic' are used throughout in normal form without diacritical notions.

3. For a full biography of Ibn `Arabî see Claud Addas, *Ibn `Arabî ou la quête du soufre rouge* (Paris, 1989); for an introduction to his life, work and thought, see S.H. Nasr, *Three Muslim Sages* (Cambridge, Harvard UP, 1964); R. W. J. Austin's introduction to *Ibn Al-`Arabî. Thc Bezels of Wisdom* (New York, Paulist Press, 1980); and A. Afifi's introduction to his commentary on *Fusûs al-Hikam* (Beirut, Dâr Al-Kitab Al-'Arabî, 1980).

4. *Al-Futûhât al-Makiyyah* ('The Meccan Revelations') is Ibn `Arabî's greatest and most renowned book. Ibn `Arabî left a large body of knowledge which, according to Ibn `Arabî himself, comprised 289 books and treatises, and according to others 400 or 500. These include many voluminous works, such as *Al-Futûhât*, the lost 95-volume interpretation of the Quran and the lost 300-chapter book *Manâhij Al-lrtiqâ'*. This paper is based on the published version of the Meccan Revelations: *Al-Futûhât al-Makiyyah fî Ma`rifat al-Asrâr al-Mâlikiyyah wa -l-Malakiyyah*, (Beirut, Dâr Sâdir, nd, 4 vols.). In all subsequent notes and captions it is referred to as *Fut.*

5. Ibn `Arabî, *Fut.*, 3:420. Indefinity is distinguished here from infinity by relating infinity to the realm of Divinity and restricting indefinity to the realm of existence. See Guénon, Réné, *The Multiple States of Being*. Trans. J. Godwin (New York, Larson Publications, 1984), 27-8.

6. The human act of bringing something into existence, that is, human creativity is the basis of this pattern. In order to be able to produce something, an

artisan must, first of all, be alive, must know what he is about to produce, which includes imagining the form of his product (imagining and knowing are, in a certain respect, identical), must have the will for production, and must have the power to produce. These are the minimum requirements without which no production is possible.

7. This sacred quadrature can also be traced back to another combination of Divine Names mentioned in the Quranic verse: "He is the First and the Last, and the Outward and the Inward" (LVII, 3), a combination whose inherent order corresponds to the order of the circle. See Ibn `Arabî, *Fut.*, 3: 197-8.

8. Ibn `Arabî, *Fut.*, 1:293.

9. Because knowledge is an inherently creative attribute in that it necessarily manifests a relationship between the knower and the known, the creator and the creature, Al-Qâshânî considers Knowledge to be nobler than Life, and consequently to be the leader (*mâm*) of the Names in the creative process. MS. 5258, f 90.

10. Ibn `Arabî, *Fut.*, 1:293.

11. Ibid., 1:293.

12. Ibid., 1:430.

13. Ibn `Arabî, *Fut*, 1:430. Ibn `Arabî explains the underlying pattern of Nature to be based on four Realities, two of which are active and two passive. Yet all are in the state of passivity with regard to the source whence they proceed. He also differentiates two levels of Nature: Grand Nature (*al-tabî`ah al-`uzmâ*), which he relates to God (*al-haqq*), and Nature which is the Principle of natural manifestation. He compares them to a mother and her daughter: the daughter is generated by her mother, but she nevertheless generates in the same way her mother does; both are capable of progenitive maternity.

14. This is based on a Prophetic tradition, see Ibn `Arabî, *Fut.*, 1:293.

15. Ibn `Arabî, *Fut.*, 1:293

16. Ibid., 1:293

17. Ibn `Arabî, *Fut.*, 4:210. "As for the action of 'creating' in the sense of the Arabic word *khalaqa*, it is synonymous with 'assigning to each thing its proper measure'." Burckhardt, Titus, *An Introduction to Sufi Doctrine*. Trans. D. M. Matheson (Wellingborough, Northamptonshire, Thorsons Publishers Ltd, 1976), 58.

18. Ibn `Arabî says that we can only say that God 'determines' (*yuqaddir*) things eternally, but not that He brings things into existence (*yûjid*) eternally, because that is impossible. Ibn `Arabî, *Fut.*, 1:388. See also *Fut.*, 4:210.

19. Ibn `Arabî says that as God knew Himself He simultaneously knew the world (*alima nafsahu fa-`alima -l-`âlam*). Ibn `Arabî, *Inshâ' al-Dawâ'ir* (Leiden, E.J. Brill, 1339 AH), 45; also *Fut*, 1:90.

20. Al-Fayrûzabâdî, *Al-Qâmûs al-Muhît* (Beirut, Mu'assaat al-Risâlah, 1986, 1st ed.), K.W.N.

21. Ibn `Arabî, *Fusûs al-Hikam*, ed. by A. Afîfî (Beirut, Dâr al-Kitab al-`Arabî, 1980) 115. See. Izutsu, Toshihiko, *Sufism and Taoism. A Comparative Study of Key Philosophical Concepts* (Tokyo, Iwanami Shoten, 1983), 198-200;

and Austin, R.W J, Ibn al-`Arabî, *The Bezels of Wisdom* (New York, Paulist Press, 1980), 141.

22. Ibn `Arabî, *Fut.*, 1:260.

23. Ibid., 1:100.

24. Ibid., 3:276.

25. See Ibn `Arabî, *Fut.*, 2:493.

26. The Sufis derive these names from the Quranic verse: "His verily is all creation and commandment" (VII, 54). See Abu Zayd, Nasr Hamid, *Falsafat al-Ta'wîl. Dirâsah fî Ta'wîl al-Qur'ân `ind Muhyî al-Dîn Ibn `Arabî,* (Beirut, Dâr al-Tanwîr li-l-Tibâ`ah wa-l-Nashr Al-Abyârî, Abd al-Hâdî, 1983) 97-149.

27. Ibn `Arabî, *Fut.*, 3:429-30. See Murata, Schiko, "The Angels," in Nasr, S. H. (ed), Islamic *Spirituality 1. Foundations* (New York, Crossroad, 1987) 330.

28. Murata, *op.cit.*, 330.

29. Ibn `Arabî, *Fut.*, 1:199.

30. The practical faculty is responsible for the manifestation of forms, while the theoretical faculty, is responsible for knowing their measures and proportions. Ibn `Arabî, *Fut.*, 2:429.

31. Also referred to as *al-habâ'* and as 'Universal Reality' (*al-haqîqah al-kulliyyah*), *Fut.*, 1:118-20.

32. Ibn `Arabî, *Fut.*, 2:431.

33. "Wholly 'formless' matter can neither be represented nor imagined, for it is a pure potentiality (that of taking on form) and has in itself no discernible characteristics whatsoever. It can only be known at all in its relationship with 'form'." Burckhardt, Titus, Alchemy, Trans. W. Stoddart (London, Element Books, 1986, 2nd. impr. 1987), 62.

34. Ibn `Arabî, *Fut.*, 3:431.

35. Ibid., 1:431. With reference to the Quranic verse, "The angels and the Spirit ascend unto Him in a Day whereof the span is fifty thousand years" (LXX, 4), Ibn KAthîr interprets the 'fifty thousand years' as the period required to traverse the spatial extension of the Throne. He says, "The distance between the Throne and the seventh earth is fifty thousand years of travel, and its breadth is fifty thousand years." *Al-Bidâyah wa -l-Nihâyah.* (Beirut, Maktabat al-Ma`ârif, 1985, 7 vols.).

36. Schuon, Frithjof, *Dimensions of Islam*, trans. P. N. Townsend (London, George Allen Unwin, 1969), 111-20.

37. Ibid., 120.

38. Ibn `Arabî, *Fut.*, 2:436.

39. Natural colours is said to have derived from the Throne, which is made of many colourful lights: green light from which comes the greenness of green, yellow light from which comes the yellowness of yellow, red light from which comes the redness of red and white light, the light of lights, from which comes the light of the day. Al-`Amidî, MS. 6102, f 8.

40. With reference to the renowned Sufi master, Ibn Masarrah, Ibn `Arabî) affirms these forms of the bearers, which correspond to the animals in Ezekiel's vision. See *Fut.*, 1:149.

41. Schuon, *op. cit.*, 109.
42. Ibn `Arabî, *Fut.*, 3:433.
43. Al-`Amilî, MS. 3103, 2.
44. The Qur'ânic references for the Gardens are:

1. 'Garden of Eden' (*jannat `adan*): "Gardens of Eden, which the Beneficent hath promised to His slaves in the Unseen" (XIX, 61).
2. 'Garden of Paradise' (*jannat al-firdaws*): "Lo! those who believe and do good works, theirs are the Gardens of Paradise for welcome, wherein they will abide, with no desire to be removed from thence" (xvm, 108-9).
3. 'Garden of Delight' (*jannat al-na`îm*): "And place me among the inheritors of the Garden of Delight" (XXVI, 85).
4. 'Garden of Abode' (*jannat al-ma`wâ*): "By the lote-tree of the utmost boundary, nigh unto which is the Garden of the Abode" (LIII, 14-15).
5. 'Garden of Immortality' (*jannat al-khuld*): "Say: 'Is that (doom) better or the Garden of Immortality which is promised unto those who ward off (evil)r' (XXV, 15).
6. The 'Abode of Peace' (*dâr al-salâm*): "For them is the abode of peace with their Lord" (VI, 128).
7. The 'Mansion of Eternity' (*dâr al-iqâmah*): "Who, of His grace, hath installed us in the mansion of eternity, where toil toucheth us not nor can weariness affect us" (XXXV, 35). In the hereafter, says Ibn `Arabî's, the concave surface of the sphere of the fixed stars will become the roof for the Fire (*al-nâr*), which will occupy the realm of the heavens and the earths.

45. Ibn `Arabî, *Fut.*, 3:435. *Al-wasîlah*, the Quranic term mentioned in (V, 35) and (XVII, 57), designates, according to Ibn `Arabî, the most superior of all Gardens, which is assigned to the Prophet Muhammad.

46. The ternary structure of the natural world refers to the three worlds in which nature reveals its governing conditions. These are the present world, the hereafter world and the intermediary world in which the souls reside awaiting the end of the present world. See Burckhardt, Titus, *Mystical Astrology According to Ibn `Arabî*, trans. B. Rauf. Aldsworth, Gloucestershire (Beshara Publications, 1977), 21.

47. The names of the Gardens derive from the Quran: "But Allah chooseth [*yakhtass*] for His mercy whom He will" (II, 105); "Such is the Garden which We cause the devout among Our bondmen to inherit" (XIX, 63); "Give thou good tidings to those who believe and do deeds of righteousness, that for them await gardens underneath which rivers flow" (II, 25); trans. Arberry, A.J., *The Koran Interpreted*. (London, George Allen & Unwin, 1983).

48. Ibn `Arabî, *Fut.*, 2:441. For more details see also 1:317-8.

49. Ibid., 2:550.

50. Ibid., 2:550.

51. Ikhwân Al-Safâ', *Rasâ'il Ikhwân al-Safâ' wa Khillân al-Wafâ'* (Beirut, Dâr Sâdir, nd., 4 vols.), 2:15; see trans. Nasr, S.H., *An Introduction to Islamic Cosmological Doctrines* (London, Thames & Hudson, 1978, rev. ed.), 65.

52. Ibid., 2:16; cf. Ibn `Arabî, *Fut.*, 2:434.

53. Ibn `Arabî, *Fut.*, 2:434.

54. Ibid., 2:433.

55. Ibn `Arabî says that the first movement is caused by the disequilibrium between the opposite natural tendencies (i.e., heat-cold, moistness-dryness), for originally nothing can manifest from equilibrium, and with equilibrium neither can there be bringing-into-existence nor production. The equilibrium was broken when heat dominated the other natural tendencies, and as the amount of heat in the body of the sphere increaased, the sphere moved. *Fut.*, 2:434.

56. According to Ibn `Arabî, the Day determined by the movement of the *Atlas* sphere in relation to the Divine Feet starts with Gemini (*Fut.*, 2:437); the day determined in relation to the sphere of the fixed stars starts with Libra (*Inshâ' al-Dawâ'ir*, 1339 AH, 61); and the day determined in relation to the sun starts with Aries (*Fut.*, 3:461).

57. Ibn `Arabî, *Fut.*, 2:438.

58. Ibid., 1:121.

59. Ibid., 2:443. The reference here is to the five stars of retrograde motion

60. It is of interest to note that *al-ahad* (=Sunday) is the same root from which derives the name of the supreme Divine state of Transcendent Unity (*al-ahadiyyah*).

61. Ibn `Arabî, *Fut.*, 2:437-38.

62. The names of the first five days of the week derive from the Arabic names of the first five numerals (1 to 5), and correspond to the five stars of retrograde motion. *Fut.*, 1:643.

63. Ibn `Arabî, *Fut.*, 1:140. In Arabic, there exists a distinction between *nahâr*, 'day-time' (of 12 hours) which extends from sunrise to sunset, and *yawm*, 'day' (of 24 hours), which extends between two successive sunrises.

64. `Abd Al-Razâq al-Qâshânî (d. 1330) is known for his commentary on Ibn `Arabî's last and most famous work *Fusûs al-Hikam*. The following is an interpretive exposition of extracts of MS 5258.

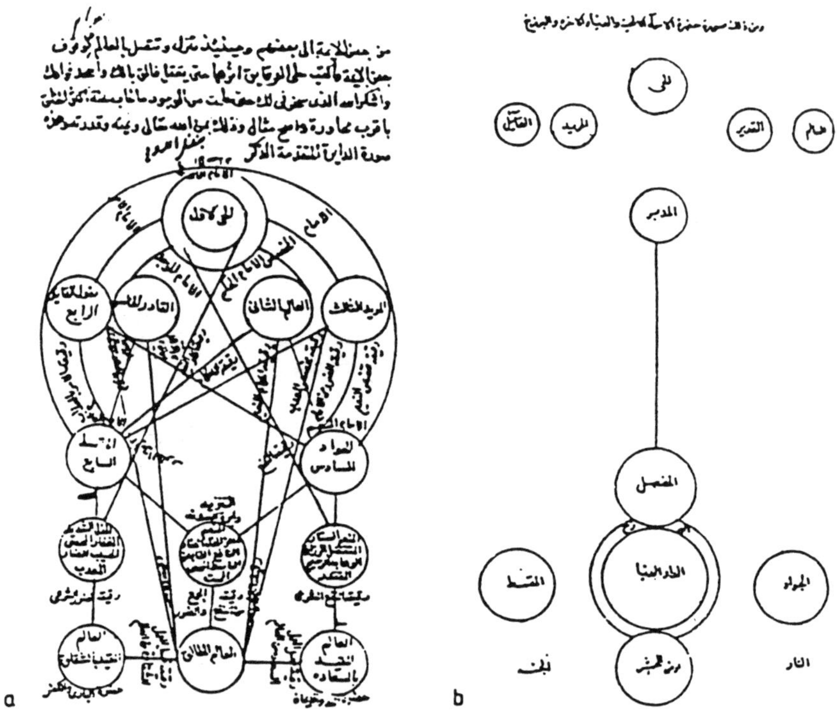

Figure 1. Two illustrations of the bounding relationships between various Divine Names. **a.** Ibn`Arabî, MS. 3975, f. 14; **b**. Ibn `Arabî, MS. 1330, f. 130.

Figure 2a. Ibn `Arabî's diagram of quaternary manifestation (MS. 1328, I, f. 127). Note that the quaternary pattern derives from the hexad, the natural division of the circumference of a circle by its radius. The hexad is based on triplicity, which represents the pattern that complements quadrature in the cosmogonic process.

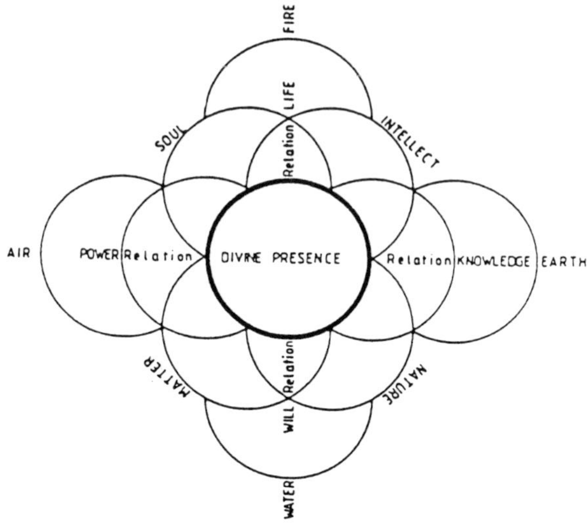

Figure 2b. English translation of 2a.

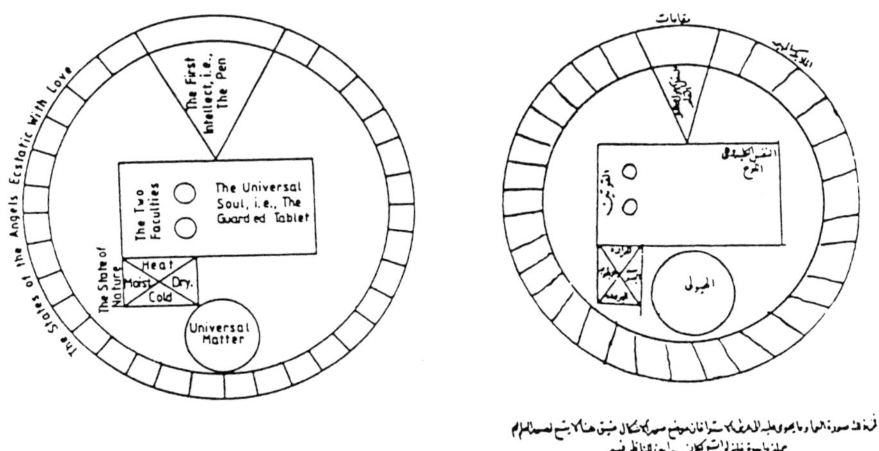

a

b

Figure 3. Ibn `Arabî's representation of the World of Command which is bounded by the circle of the primordial 'Cloud' (*`amâ')*, its higher limit, and the 'Resting Throne' (*`arsh al-istiwâ'*), its lower limit. The Throne is contained in the circle of 'Universal Matter' (*al-hayûlâ al-kull*). (**a:** Ibn `Arabî's, *Fut.*, 3:421; **b:** MS. 1330).

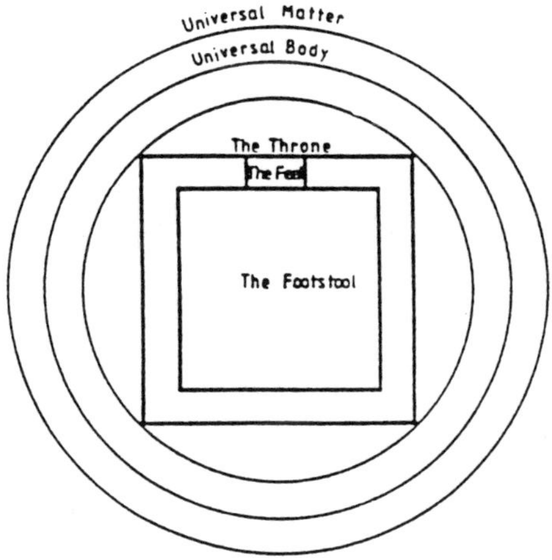

Figure 4. Ibn `Arabî's representation of the 'Resting Throne' (*`arsh al-istiwâ'*), the Footstool (*al-kursî*) and the Feet (*al-qadamân*). (Ibn `Arabî, *Fut.*)

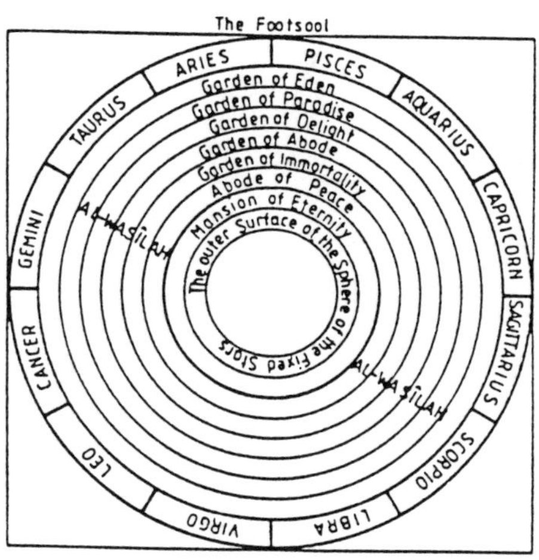

Figure 5. Ibn `Arabî's representation of the celestial Gardens, showing the 'sphere without stars', the Gardens, the outer surface of the 'sphere of the fixed stars'. (Ibn `Arabî, *Fut.*).

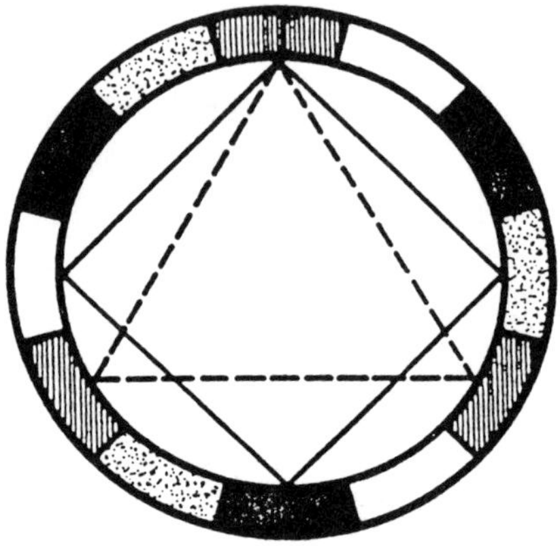

Figure 6. The duodenary pattern of the zodiac generated by the trigon and the square (the shadings differentiate the natures of the signs).

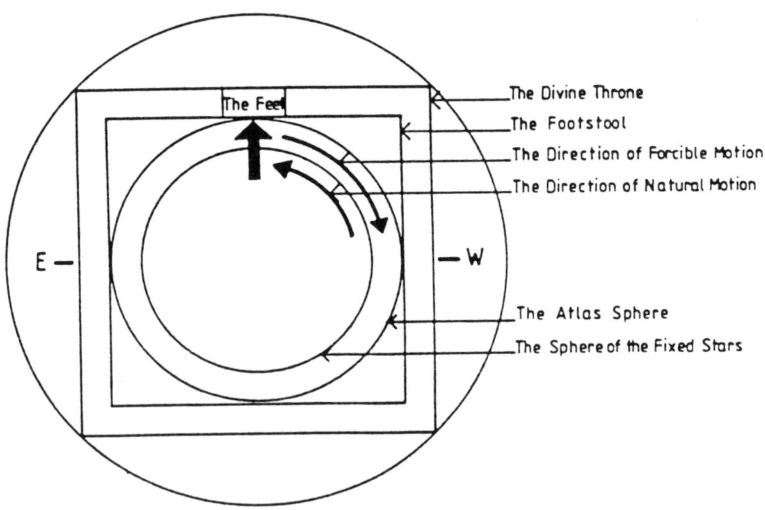

Figure 7. The relation of the *atlas* sphere to the Divine Feet.

The Use ot Synchronic Contingency in Early Fourteenth Century Debate over the World's Temporal Duration

Ian Wilks

There is a key temporal dimension to medieval conceptions of contingency; recent scholarship has, in particular, suggested that an important theoretical divide exists in the way medieval theorists establish a defining temporal reference point for the contingent. In certain discussions concerning the possible eternity of the world, for instance, we encounter these rival conceptions:

(i) Something is contingent if it could have become presently otherwise than it is (i.e., if a different past course of development could have left it in a different present state). Contingency resides in a past possibility.

(ii) Something is contingent if it can be presently otherwise than it is (i.e., if a presently unrealized possibility can still be considered a present possibility, even though unrealized). Contingency resides in a present possibility.[1]

These conceptions are different, but not inconsistent; one can readily hold that contingency holds in both respects. But grounds for disagreement exist where one party appeals only to (i), and refuses any appeal to (ii), in way of explaining contingency, while another party maintains the reasonableness of appealing to (ii). Cases can exist where rejecting or accepting (ii) can make considerable difference in the understanding of some issue; there is, indeed, one particular aspect of the eternity of the world debate which clearly constitutes such a case.

It is easy to see how either employing (i) or (ii) will shape our understanding of contingent statements. Take, for example, the statement "Socrates is sitting." (i) tells us that its contingency lies in this fact: while sitting now, Socrates did have the ability to bring it about that he was standing now, not sitting. But (ii) tells us the contingency lies in this fact: while sitting now, there still exists for Socrates the unrealized possibility that he could be standing now; the sitting-alternative and the standing-alternative are both possibilities for the present, one of which is actually realized, the other not. But according to (i), only the sitting-alternative -- the realized alternative -- is a present possibility. The standing-alternative is ruled out as a possibility for the present. There was a past possibility for its holding in the present, but given that it does not now hold, there is no present possibility for its holding in the present.

(ii), with its positing of simultaneous contrary possibilities, is referred to, in an obvious way, as "synchronic contingency," while (i), where the contrary possibilities are understood as successive, is called "diachronic contingency." Of the two, the diachronic approach seems more intuitive. Since it is impossible to sit and stand at the same time, one naturally supposes that when the possibility of doing the first is realized in the present, the possibility of doing the second is automatically removed for the present. The proponent of the synchronic approach argues a more challenging case. If two alternatives are contrary, and one is now realized, the other is still to be considered now possible. But does not the realization of one of those contrary possibilities rule the other possibility out? The answer, in effect, seems to be that it rules out the realization of the other possibility, but does not rule out the possibility itself. Contrary possibilities can arise simultaneously, and either one can remain as a possibility even as the other is realized.

As a matter of historical development, it seems that the diachronic approach is older and more perduring in the medieval tradition, and that a full-fledged synchronic approach emerges largely in the fourteenth century, with some minor presence in earlier eras.[2] One issue which seems to have encouraged this shift in modal theory was the continuing debate over free will. The will, argues Scotus (and before him, Olivi), possesses potencies for contrary alternatives -- in this case, contrary acts.[3] These potencies are not only able to be successively present but simultaneously present as well, and the contingency of an act of will thus lies in the accompanying and unrealized potency for a contrary act of willing. The freedom of the will is thus guaranteed, and indeed is presented in the strongest sense; we are free not just because we can will otherwise later, or could have willed otherwise before, but because we can will otherwise now, at the very instant when the contrary potency of the will is realized and fully in play. The synchronic interpretation of the will's contingency thus informs an especially pronounced libertarianism.

II

Of course it is not only humans who have free will; God has it, presumably, as well, and manifests it in the free, creative activity he undertook to bring about the world of creatures. It is in this connection that another -- and as yet less examined -- application of the synchronic/ diachronic modal distinction arises in medieval debates. It would seem that if the world God created were of unlimited temporal duration extending into the past -- that is, if it were without beginning in time -- then a question may be raised as to how God's act of creation could be other than a necessitated one, and thus how the divine willing underlying that act could actually be free. Henry of Ghent, in his famous *Quodlibet* 1.7-8,[4] raises precisely this question, and answers negatively, that the act could not be other than necessitated, that the willing could not be free; the unacceptability of these results is proof, for him, that the world is not of unlimited temporal duration extending into the past.

His argument explicitly presupposes a diachronic approach to modality; this is immediately evident when he draws a distinction between two types of necessity -- absolute and non-absolute -- in terms of the Aristotelian dictum (as I shall tend to call it) that whatever is, when it is, necessarily is.[5] What this dictum means, of course, is that while something presently is, no contrary alternative to it is concurrently possible; that alternative may have been previously possible, or may be subsequently so, but not concurrently. Now this account is at its core diachronic, since a synchronic account would readily allow the contrary alternative to be possible in the present, putting emphasis on the fact, rather, that it is a presently unrealized possibility. So the ensuing argument which Henry provides depends on an diachronic account, inimical to its synchronic counterpart -- and this is a fact which will be used against the argument by Scotus.

The Aristotelian dictum actually defines non-absolute necessity; something is non-absolutely necessary in this sense just by virtue of being presently as it is, and so ruling out as impossible -- for the present -- all contrary alternatives. On Henry's formulation, there is no "potency for a contrary at that time when the act is posited"; rather, potency for a contrary is a "potency posited in being at some other time, in which the act is able to be impeded."[6] To claim, as is implied here, that anything which presently is, is necessary in this non-absolute sense, is not very contentious; having this necessity in the present is no guarantee of having it in the past or future. But if there *were* this guarantee of having it at all times, past, present and future, then what would be achieved would be the much stronger sense of necessity Henry refers to: absolute necessity.

Imagine that the world is eternal. Each moment it exists it is non-absolutely necessary, meaning that, at each moment, it has no potency for non-existence. But because it exists at *all* moments of time -- being eternal -- it follows that at all moments it has no potency for non-existence. This would seem to be another way of saying that it can never not exist, hence that it is necessary. But this is a special, enhanced necessity which we now see attaches to its existence, an *absolute* necessity, as Henry puts it -- which is, in effect, a non-absolute necessity that applies through every moment of time. So we say, in conclusion, that an eternal world will be an absolutely necessary world.[7] "Thus if the created world is taken as always having had being from God and from eternity then it is absolutely necessary for it to have been always and from eternity, and if it is so ... then there was never any potency by means of which it could at some time not have been."[8]

Say the world's eternity *does* imply its necessity in this way. But its necessity would seem, in turn, to imply a certain divine inefficacy in the act of creation, since God evidently has no option *not* to create this necessary world. Those who maintain the eternity of the world, then, in effect maintain that it "has its being from God only by necessity of nature, or by a will which is conjoined with the necessity of an immutable nature, and which is not free to give, or not give, being."[9] So God would seem not to have chosen freely between the options of imparting being to the world or not, and in this way, Henry argues, the world's eternity implies a certain constraint upon the freedom of God -- an implication that is, in an obvious way, theologically unpalatable.[10]

III

The upshot of this analysis is to reject the position that the world is eternal, and thus actually to prove that it was created a finite time ago. The line of attack here is highly novel, and Richard Dales notes that he has "never come across this argument in medieval discussions of the eternity of the world, and indeed Henry seems to have invented it himself, according to his understanding of Aristotle and Avicenna."[11] Indeed, the argument represents a distinct reversal in the way reference to God's will figures in the debate over the world's longevity. It had traditionally seemed more difficult to square the world's non-eternity with a theologically acceptable account of the divine will -- the issue being that if the world had been created after a period of non-existence, this could only come about through an alteration in God's will, an alteration whose very occurrence would then suggest that God's will was mutable.[12] A non-eternal world thus implies a mutable divine will; whereas, for Henry, an eternal world suggests an unfree divine will. No doubt because the former inference had already been circulating in medieval debate, giving rise to various documented refutations, while Henry's was new and unusual, it seems to have supplanted the other.

Where previously the burden of proof is with those who oppose the possibility of an eternal world to show their view compatible with accepted views on the divine will, now it shifts to those who accept that possibility to demonstrate compatibility.[13]

Of note among the latter group are Scotus and William of Ockham. Scotus, as noted, champions synchronic contingency (and may well have been its chief medieval innovator); Ockham opposes it, and cleaves to a traditional, diachronic model. This, in the end, as we shall see (and as it is the chief purpose of this study to show), leaves Ockham in a very difficult position in his attempt to refute Henry, and involves him in perplexities as great as those which he endeavours to argue away.

The contrast with Scotus in this regard is marked. Henry's argument -- with its diachronic orientation -- is almost automatically disarmed by a synchronic understanding of contingency, such as is advanced by Scotus. As will be recalled, Henry stipulates that there is no "potency for a contrary at that time when the act is posited,"[14] and uses this to confirm his interpretation of the Aristotelian dictum as asserting the necessity of the present. But Scotus, of course, maintains that there *is* potency for the contrary at the time when the act is posited, and accordingly can do no other than reject this doctrine of the necessity of the present; for him, the contingency of the presently realized alternative is able to be guaranteed by the unrealized, contrary alternatives that lurk simultaneously behind it. Now if the Aristotelian dictum could only be read in the way Henry suggests -- as affirming the necessity of the present -- Scotus would thus have to reject it. But what he does instead is re-interpret it. What it affirms, says Scotus, is not the necessity of a thing's existence at the time when it exists, but rather the necessity of this statement: "Everything is when it is," which is self-evidently true.[15] On this reading the necessity is safely transferred from the present thing -- where it is metaphysically problematic -- to a tautology about the present thing, which is necessary in an obvious and metaphysically innocuous way (we shall shortly examine this strategy a little more fully, as it unfolds in the hands of Ockham); hence Aristotle is not saying that a present thing is presently necessary, and so is not denying that a present thing can be presently contingent. When, for example, the present thing is the will's present act of willing, the dictum is perfectly consistent with the position that "the will, when desiring something at A, is able not to will it absolutely at A,"[16] and so, Scotus concludes, "in no way is necessity of concomitance signified"[17] -- in other words, one's act of willing something does not go together with a concomitant necessity of willing that thing. Rather, it goes together with a concomitant, but unrealized possibility (i.e., potency) for willing something different.

Scotus comments more extensively upon the interpretation of this dictum as it bears upon free will than as it bears upon the eternity of the world; in the latter case, in fact, he does very little else than simply refer us back to his discussion of the former,[18] essentially pausing only to extract this precept applicable to a non-necessary being: "In that instant and for that instant in which and for which it is, it contingently is."[19] This, of course, is of itself quite enough to dispense with Henry's argument. But the reference to the material on willing tells us more. Just as we can say of a human will that what it wills at A it can also not will at A, so we can say that "the divine will, willing at the instant of eternity that a stone exist, is able at the same instant to will that the stone not exist."[20] Thus, again against Henry, the divine option of creating or not creating can exist even with respect to an eternal creature. This is because a cause need precede its effect only *in nature*, not necessarily *in duration* -- as we find (to use the best-known instances of such causes) in the case of the sun, which produces its rays as *soon* as it exists,[21] and the footstep, which produces its footprint as soon as it steps. Simultaneous causes are, by definition, not prior in duration, but as causes, they must still be identified as being in some sense prior to their effects, and hence are called prior in nature. Relative to the world God has that kind of priority, and so can exercise full freedom of creation even over an eternal world; the fact that one of two synchronic options -- creating -- was exercised at the instant of eternity, is no bar to fact that the other -- not-creating -- was likewise available to God, though unexercised, at that same moment of eternity.

IV

The core of Henry of Ghent's position is to distinguish two grades or levels of necessity: the necessity conferred upon a present thing by the truth of Aristotle's dictum ("Everything, when it is exists, necessarily exists"), and what Henry calls "absolute necessity." To be necessary in the first, weaker sense, is metaphysically trivial; since everything achieves this necessity, it fails to mark out any special class of beings.[22] To be necessary in the second, stronger sense -- to be absolutely necessary -- is of enormous metaphysical import; it is to be a necessary being, as that phrase is normally understood. The gap between these two necessities is very great, but, as Henry in effect argues, is it capable of being bridged. Something which is *always* necessary in the weaker sense is established automatically, by that fact alone, as necessary in the stronger. This is the connection which informs Henry's basic argument: that if something is an eternal being then it is a necessary being as well.

Opponents of this argument want to show that this gap between the weaker and stronger senses of necessity is incapable of being bridged. Scotus, in fact, demonstrates this quite easily, given his assumption that contingency can be synchronic; the necessity expressed in the Aristotelian dictum is merely the necessity of the statement "Everything, when it is, is," and not the present necessity of every existing thing. This holds, as we have seen, because corresponding to realized potencies are simultaneous, unrealized, contrary potencies. The result is that things are shown not to manifest the weaker sense of necessity invoked by Henry -- that, in other words, the Aristotelian dictum expresses a sense of necessity *even weaker* than that which Henry attributes to it. Hence the gap between it and the stronger, absolute sense required for Henry's argument yawns even wider, and is correspondingly less traversable.[23]

Now William of Ockham must deal with this same argument of Henry's, since he too is inclined to consider God to have been capable of producing an eternal world;[24] but on the matter of the nature of contingency he is resolutely opposed to Scotus -- his account of modalities is, as I have noted, firmly diachronic in orientation. So he must refute Henry's argument using some of Henry's own diachronic modal assumptions -- and in consequence, as one can readily see, his refutation does not fare nearly so well.

He begins, as Scotus does, by rejecting Henry's interpretation of the Aristotelian dictum, and in much the same way, too; "Everything, when it is, necessarily is" predicates necessity of a statement, Ockham explains, not of the existence of particular things. The dictum simply tells us that this statement is necessarily true: "Everything that is, is when it is"; it qualifies the *whole statement* as necessary.[25] It does not simply attach to one of the clauses, as in: "Everything that is, necessarily is when it is." The point becomes a little more evident when we consider instances of these general principles (and thus forgo the complicating presence of the quantifier). On Ockham's reading, the dictum implies this about Socrates: "Socrates is when he is" is necessarily true. By contrast, on his reading the dictum does not imply this: "Socrates necessarily is when he is." Why is the latter to be rejected? He points out that it has the form of a temporal statement, where the subordinate clause stands in some clearly expressed temporal relation to the main clause -- expressed in this case by the conjunction "when." This statement about Socrates has the form "X when Y," where "Socrates necessarily is" substitutes for "X" and "He is" for "Y." But statements of the form "X when Y" require both X to be true and Y to be true (indeed, they require also that X and Y be true at the same time);[26] thus if "Socrates necessarily is when he is" is true then "Socrates necessarily is" is true as well -- and, of course, this latter statement is manifestly false. So if we read the Aristotelian dictum in such a way as to make it imply "Socrates necessarily is when he is," then we end up attributing necessity to individual, categorical components of "X when Y"-statements in a totally inappropriate and erroneous way.

The lesson -- presented more succinctly by Scotus -- is that the claim of necessity must attach to the whole statement, not to the parts; that "Socrates is when he is" is necessary taken as a whole, but exhibits no necessity in respect of its parts (particularly in respect of its first part, Socrates is"); and that the Aristotelian dictum must be read in a way consistent with these insights. This is enough to impede the first step in Henry of Ghent's argument, a step which depends upon the very interpretation of the dictum which Ockham successfully rejects; we now see that Henry cannot appeal to the dictum to prove that present existence implies present necessity. We would thus seem to have no reason to follow Henry in concluding that something existing for all time is necessary in an absolute sense.

But as it happens, Ockham does seem to have additional trouble dispensing with this latter stage of Henry's argument, even though he successfully refutes the former; the problem lies in the fact that Ockham is perfectly willing to concur with Henry's diachronic approach to contingency. The relevant line of argument is this. The fact that Socrates' existence at a given time (call that time "A") is contingent may be explained synchronically by his having at A a potency for non-existence at A, or diachronically by his having prior to A a potency for nonexistence at A. Ockham, consistent with his resistance to Scotus' views on synchronicity, clearly rejects the former option for the latter: "I concede that Socrates was able not to have existed at A. And when you ask 'whether he was therefore able, at A, not to have existed at A, or prior to A,' I say that it was prior to A."[27] So, in short, the contingency of the existence of anything lies in a prior capacity not to have been; where this prior capacity is lacking, the thing is not contingent.

Now this account of contingency seems, in an obvious way, to prevent any eternal thing from qualifying as contingent, since for an eternal thing there can be no prior capacity for non-existence; this last statement holds because, by definition, there can be nothing prior to an eternal thing. By what I have just stated in effect amounts to a precis of Henry's argument; and the point is that refuting that argument is not just a matter of rejecting his reading of the Aristotelian dictum, but, it seems, is also a matter of rejecting the whole diachronic approach to con-tingency which informs that reading (and, indeed, it may be the chief purpose of the reading to convey, or even justify, that diachronic approach[28]). So Ockham may question Henry's argument in a more superficial way by questioning his reading of the Aristotelian dictum, but this seems in danger of being to no avail if the whole diachronic modal framework is left intact. If the contingency of a thing's existing requires a prior possibility of non-existing, and yet there are no prior possibilities for an eternal thing, how does eternity *not* imply necessity? How does Ockham avoid admitting that an eternal world would be a necessary world?

Surprisingly -- given his opposition to Henry -- he actually makes the admission that an eternal world would be necessary, and merely argues that the necessity so attaching to it would be of a lesser sort, not at all like something we could consider an *absolute* necessity. On the assumption that the world has always existed, Ockham says, "this proposition 'the world was from eternity' is now necessary, since this is now true and never was able to be false. However it was able to have been false just as any true proposition about the past is necessary, and nonetheless was able to have been false."[29] What is referred to here is often called "historical necessity." Caesar had the historical option of either crossing or not crossing the Rubicon, but once he *did* cross, the fact of his doing so -- previously undetermined -- became historically fixed, and could not subsequently be reversed. Anything that happened in the past is in this way now necessary -- but, in a sense it did not *have* to be necessary, because some alternative could have happened in the past in place of it. So this is undoubtedly a weaker sort of necessity than, say, the necessity of God's being good or wise,[30] characteristics which are intrinsically necessitated by the very nature of God; the form in which the past was fixed was, by contrast, *not* intrinsically necessitated (since it could have been otherwise), and so seems to constitute a kind of necessity falling short of this absolute degree. *This* is the degree of necessity Ockham attributes to an eternal world; it may be necessary, but in some sense did not have to be necessary, because, presumably, God could have chosen it to be otherwise.

But it is not clear how this account will work for an eternal object, since there is no preceding interval of time during which its coming to be is under question. *How* could God have chosen it to be otherwise, given that it always was? Ockham acknowledges this objection. Assuming that God created the world from eternity, it seems that "he was not able not to have produced the world, because <he was neither able not to do so> before eternity, since nothing is before, nor <was he able not to do so> in eternity owing to <what would otherwise be> a contradiction; therefore he produced <the world> necessarily."[31] God's option to create or not create did not exist *in* eternity because it would then be contemporaneous with the eternal word itself, and this, given Ockham's rejection of any synchronic account of contingency, would be anomalous -- a contradiction, indeed. But neither did his option to create or not create exist *before* the eternal world because, obviously, nothing can be before something which is eternal. Hence, it would seem, God had no option at all, and if he produced an eternal world, he produced a necessary one, and produced it necessarily as well.

So in spite of his considerable disagreement with Henry's interpretation of the Aristotelian dictum, and his extended critique of that interpretation, Ockham seems to accept, in the end, much of Henry's ensuing argument. The following inference is representative, in a schematic way, of Henry's view:

[1] An eternal creature is a necessary creature
[2] A necessary creature implies limits to divine power
[3] Therefore an eternal creature implies limits to divine power (i.e., the unlimitedness of divine power implies the non-eternity of all creatures).

Scotus successfully blocks [1]; Ockham appears to be attempting to do so for much of his argument, but in the end seems to reconcile himself to [1]. His argument concludes with a very brief, and last-ditch, attempt to block [2], which consists mostly of the following (where "A" denotes eternity):

> But if we should hold that these stand at the same time -- that the world was contingently produced by God, and from eternity -- then on this basis we must say, as it seems, that the world was able not to have been at A, since God was previously, by nature or *secundum intellectum*, able not to have produced the world at A, since it was in his will to produce <the world> at A or not to produce it.[32]

Ockham here invokes -- as we have already seen Scotus do -- priority of nature to provide the requisite defence of divine power. The will of God acts at an instant not prior in time but prior in nature to the world; *this* priority applies just as readily to an eternal world as to one of finite duration, and suffices to attach the right degree of contingency to its creation.[33] At a stroke the problem is solved. We can now say that prior to the existence of the world, even if eternal, God could have *not* produced it, and so in producing it acted freely. But notice the deeper impasse Ockham has now arrived at by these means. He adopts the language of priority of nature, and secures the contingency of divine creation; but he retains a diachronic approach to modality, on the basis of which an eternal world is, by the fact of its eternity alone, necessary. So if the world *were to be* eternal (and Ockham is defending the possibility of its being so) then it would follow that God freely created a necessary thing -- and this seems at least perplexing, at worst a contradiction. How can a necessary thing be subject to non-necessary production and still be necessary? How can God avoid bringing about something which, having been brought about, could not have avoided existing? This seems as much as to say that God can withhold an act which, once performed, God could not have withheld. It is not, in short, clear that Ockham has reached a tenable conclusion; he may well have substituted a greater problem for a lesser.

V

What emerges from this study is the efficacy of the synchronic account of contingency, and the comparative inefficacy of the diachronic account, for dealing with the problem of divine power in the face of eternal creation.[34] The synchronic approach of Scotus attaches a great deal of sense, and some rigour, to the possibility of free, eternal, divine creation, while the diachronic account of Ockham fumbles with this possibility, and in the end maintains it at the cost of advancing what seems to be an absurdity. The superiority of the Scotistic handling of this issue may thus viewed as further evidence for the philosophical significance of the developing, largely fourteenth century interest in synchronic contingency.[35]

Erindale College, Univeristy of Toronto in Mississauga

Notes

1. Simo Knuuttila has been chiefly responsible for the growing scholarly fascination with the second of these accounts of contingency ("synchronic contingency"), and for promoting the view that it represents a significant, largely fourteenth-century departure from received accounts of modality. See his *Modalities in Medieval Philosophy* (London and New York: Routledge, 1993), 31-38, 143-44. Earlier works from this author which pursue the same theme include "Time and Modality in Scholasticism," in S. Knuuttila, ed., *Reforging the Great Chain of Being: Studies of the History of Modal Theories* (Dordrecht: Reidel, 1981), 163-257; and "Modal Logic" in Norman Kretzmann, Anthony Kenny and Jan Pinborg, eds., *The Cambridge History of Later Medieval Philosophy* (Cambridge, England: Cambridge University Press, 1982), 342-57. For a critical analysis of some aspects of Knuuttila's approach see Klaus Jacobi, "Statements about

Events: Modal and Tense Analysis in Medieval Logic," *Vivarium* 21 (1983): 175-89.

2. Knuuttila, *Modalities in Medieval Philosophy,* vii. Knuuttila, ibid., 45, 75-96, finds an incipient account of synchronic contingency in Gilbert of Poitiers and Peter Abelard. Neil Lewis argues for the same in Robert Grosseteste; see "Power and Contingency Robert Grosseteste and Duns Scotus," in Ludger Honnefelder, Rega Wood and Mechthild Dreyer, eds., *John Duns Scotus: Metaphysics and Ethics* (Leiden and New York: E. J. Brill, 1996) (hereafter, "Duns Scotus"), 205-25. Stephen D. Dumont likewise finds a direct inheritance from Peter John Olivi in the Scotistic account of synchronic contingency as applied to the human will; see "The Origin of Scotus's Theory of Synchronic Contingency," *The Modern Schoolman* 72 (1995): 149-67.

3. The key texts are *Lectura* 1.39, in *Ioannis Duns Scoti Opera omnia ... praeside Carolo Balic* (Vatican City: Typis Polyglottis Vaticanis, 1950-) 17:481-510; and *Ordinatio* 1.39 Apograph, in *Opera omnia* 6:417-19. For extensive commentary on the former source see Antonie Vos, trans., *John Duns Scotus: Contingency and Freedom: Lectura I 39* (Dordrecht and Boston: Kluwer, 1994).

4. Henry of Ghent, *Quadlibet* 1, ed. Raymond Macken, in *Henrici de Gandavo Opera omnia* (Leiden: E.J. Brill, 1979-) 5:27-46.

5. *Quodlibet* 1, 40 (20-21). For the Aristotelian source see *De interpretatione translatio Boethii*, ed. Laurentius Minio-Paluello, *Aristoteles Latinus* 2.1 (Bruges and Paris: Desclée de Brouwer, 1965), 17 (3-4): "Igitur esse quod est quando est, et non esse quod non est quando non est, necesse est."

6. *Quodlibet* 1, 41 (1, 2-3). The entire sentence (ibid., 40-41 [26-30]) reads thus: "Ita quod in nullo istorum modorum est potentia ad contrarium pro eodem tempore quo ponitur actus, sed si sit potentia ad contrarium, hoc est per potentiam positam in esse pro alio tempore in quo potest actus impediri, quia contingens est."

7. I am using "eternal" here to mean "of infinite temporal extent"; this is, of course, not the deepest sense attached to the word in the medieval tradition -- for which see Eleanore Stump and Norman Kretzmann, "Eternity," *Journal of Philosophy* 78 (1981): 429-58.

8. *Quodlibet* 1, 41 (48-50, 51). In Henry's formulations at this point in the argument, reference is made only to the absence of past potencies: "Si ergo aliquid aliquando fuit nec umquam erat potentia praecedens per quam actus essendi eius pro tempore quo fuit, potuit impediri, absolute necessario fuit, quia non erat omnino potentia, neque rei existentis neque causae efficientis, per quam potuit impediri ne tunc fuisset." For fuller demonstration of absolute necessity, a parallel statement for *potentia succedens* probably ought to be supplied as well.

9. *Quodlibet* 1, 43 (89-91): "Mundus habet esse a Deo sola naturae necessitate aut voluntate immutabilitatis coniuncta necessitati naturae, non libera ad dare esse et non dare."

10. For further discussion of this argument see Richard C. Dales, *Medieval Discussions of the Eternity of the World* (Leiden: E.J. Brill, 1990), 168-70 (hereafter *Medieval Discussions*); and Raymond Macken, "La temporalité radi-

cale de la créature selon Henri de Gand," *Recherches de théologie ancienne et médiévale* 38 (1971): 238-42.

11. Dales, *Medieval Discussions*, 168.

12. Aquinas reviews this argument in some detail, formulating its key insight thus: "When anyone wants to do something, and it is not immediately done, it must be the case either through a lack of power ... or because the will to do it is not yet complete" (*Summa contra gentiles* 2.32). So if God willed the world's creation eternally, but enacted the will only subsequently, how do we explain this without attributing change, and thus imperfection, to him? (Aquinas' response [ibid. 2.35] is to maintain that God wills not only a thing's existence, but the time it comes to exist as well -- and thus can quite readily will it to enter existence at any given subsequent time.) Compare this contemporary, and rather more direct, formulation of the argument, from Peter of Tarentaise, which I cite in full: "Deus aut est agens per voluntatem, aut per necessitatem naturae. Si per necessitatem: sed posita illius cause ponitur eius effectus; ergo mundus fuit ab aeterno. Si per voluntatem: sed omnis voluntas quae incipit agere de novo, variantur, quia, vel removetur aliquod impedimentum prius, vel subiectum aliqua immaginatione excitatur. Ergo, cum Dei voluntas sit invariabilis, non incipit agere de novo; ergo, egit ab aeterno" ("Circa Petri de Tarantasia quaestionem 'De aeternitate mundi,'"ed. Omar Argerami, *Patristica et Medievalia* 2 [1981], 80; for the corresponding solution see ibid., 84).

13. A form of Henry's argument exists which uses a different kind of diachronic definition of contingency. We find an example in Thomas of Wylton's quaestio, "An ista simul stent, quod motus sit aeternaliter a deo productus et cum hoc quod deus sic producit mundum libere, quod potuit ipsum non produxisse," in *Quaestiones quodlibetales*, ed. W. Senko, *Studia Mediewistyczne* 5 (1964), 162 (10-12): "Potentia liberi arbitrii non est respectu praeteriti nec praesentis, sed respectu futuri contingentis; sed si mundus ab aeterno fuisset, semper suum esse fuit praesens et praeteritum et per consequens respectu eius non fuisset nulla fuit praeteritas liberi arbitrii." Contingency is here defined not in terms of a prior possibility but a subsequent possibility; my sitting now is contingent, not because (as Henry would say) at a prior time I could have brought it about that I am now standing, but because at this present time there is a future possibility that I *will* be standing. Contingency, on this account, is future contingency. For prior sources on this view see ibid. (13-14), where Hugh of St. Victor and Peter Lombard are mentioned; see (respectively) *De sacramentis* 1.5.22, *PL* 176, 256; and, as well, *Magistri Petri Lombardi ... Sententiae in IV libris distinctae* 1.2 (Rome: Editiones Collegii S. Bonaventurae, 1971), 461 (14-15).

14. See n. 6 above.

15. *Lectura* 1.39, no. 58 (15): "Denotatur esse necessarium 'omne esse quando est.' "

16. Ibid. (20-21): "Voluntas volens aliquid in *a*, potest nolle absolute in *a*." "A" marks an instant of time (I use the letter in upper case rather than italic [as appears in the edition] for my translations to emphasize its difference from "a" taken as indefinite article).

17. Ibid. (21-22): "Non significatur aliqua necessitas concomitantiae."

18. *Ordinatio* 2.1.3, no. 157, 79 (7-9): "Ad secundam rationem, de Philosopho II *Perihermeneias* ('Omne quod est, quando est, necesse est esse'), -- patet responsio distinctione 41 primi libri." Of course the actual distinction is 39, not 41.

19. Ibid., no. 158, 79 (12-13): "In illo instanti et pro illo instanti, in quo est et pro quo est, contingenter est."

20. *Lectura* 1.39, no. 53, 497 (22-24): "Voluntas divine volens in instanti aeternitatis lapidem esse, potest nolle in eodem instanti lapidem esse."

21. Calvin G. Normore, "Scotus, Modality, Instants of Nature and the Contingency of the Present," in *Duns Scotus*, 168, summarizes this example well: "In natural philosophy the picture of light being propagated instantaneously through a diaphaneous medium by the sun ... was a clear case of a causal process in which the effect and the cause were coincident in time. Despite this temporal coincidence there is a clear sense in which the production of light by the sun is prior to its reception on the earth." S. Dumont, "Time, Contradiction and Freedom of the Will in the Late Thirteenth Century," in *Documenti e studi sulla tradizione filosofica medievale* 3 (1992), 568, comments thus on priority in nature: "That two things together in time (*simul tempore*) can be ordered so that one is prior to the other in nature (*prior natura*) was a scholastic commonplace having its root in Aristotle" (see ibid., n. 20).

22. Norman Kretzmann, "Ockham and the Creation of the Beginningless World," *Franciscan Studies* 45 (1985): 28 (hereafter, "Ockham and Creation), refers to this weaker sense quite appropriately as "conditional necessity."

23. Steven P. Marrone, "Revisiting Duns Scotus and Henry of Ghent on Modality," in *Duns Scotus*, 175-89, argues that differences between Scotus and Henry on the matter of how contingency is to be understood are not always completely clear-cut; the degree of difference between them manifested on the issue of the world's eternity is thus not necessarily present to the same degree in all other areas of dispute.

24. Useful accounts of Ockham's response to Henry are found in Norman Kretzmann, "Ockham and Creation," 24-31; and Dales, *Medieval Discussions*, 226-28.

25. William of Ockham, *Quaestiones variae*, eds. G. I. Etzkorn, F.E. Kelley, and Joseph C. Wey, q.3, in *Guillelmi de Ockham Opera philosophica et theologica* (St. Bonaventure, N.Y.: St. Bonaventure University, 1974-88) 8:87 (500-2): "Sed Philosophus dicit quod ista propositio est necessaria 'omne quod est, est quando est' quia haec non potest esse false."

26. Ibid. (506-9): "Haec enim est false 'Sortem quando est necesse est esse,' quia ad hoc quod temporalis sit vera vel necessaria, requiritur quod utraque pars sit vera vel necessaria pro eodem tempore." The logic of temporal statements is briefly discussed at ibid., 88 (528-31).

27. Ibid., 91 (579-81): "Et concedo quod Sortes potuit non fuisse in *a*. Et quando quaeris 'aut igitur in *a* potuit non fuisse in *a*, aut ante *a*,' dico quod ante *a*."

28. After all, if the present is necessary, contingency *can* only be understood with reference past or future time -- i.e., diachronically; and if contingency is synchronic then the present cannot be thought of as necessary.

29. Ibid., 91 (582-86): "Haec propositio nunc est necessaria 'mundus fuit ab aetemo,' quia haec nunc esset vera et numquam potest esse falsa."

30. Ockham uses "Deus est bonus" and "Deus est sapiens" at ibid., 89 (551-52) as examples of categorically necessary statements (as distinct from the conditional, causal and temporal statements he has just finished distinguishing on 87-88 [510-31]).

31. Ibid., 91 (591-94): "Non potuit non produxisse mundum quia nec ante aeternum quia nihil est ante, nec in aeterno propter contradictionem, igitur necessario produxit." Ockham affirms this position again at 92 (601-6): "Et ad quaesitum 'aut in *a* habet potentiam ad non-essendum' aut ante *a*,' respondeo quod si per *a* demonstres aliquod instans temporis, tunc dico quod ante *a* potuit non fuisse in *a*. Si per *a* demonstres aeternitatem, tunc dico quod nec in *a* potuit mundus non fuisse in *a* propter contradictionem, nec ante *a* quia nihil est ante *a* realiter."

32. Ibid., 92 (606-11): "Sed si teneamus quod ista stant simul quod mundus sit contingenter productus a Deo et ab aeterno, tunc oportet dicere, ut videtur, quod propter hoc potuit mundus non fuisse in *a*, quia prius nature vel secundum intellectum potuit Deus non produxisse mundum in *a*, quia in voluntate sue fuit producere in *a* vel non producere." Kretzmann, "Ockham and Creation," 29, translates the penultimate clause thus: "By natural or conceptual priority [as distinct from temporal priority] God could have not produced the world at A."

33. For the notion of priority of nature see n. 21above.

34. Scott MacDonald, in "Synchronic Contingency, Instants of Nature, and Libertarian Freedom: Comments on 'The Background of Scotus' s Theory of Will,' " *The Modern Schoolman* 72 (1995): 170, asks, "So what's *important* about the development of a synchronic view of contingency?" He wishes to suggest, in essence, that not very much is. He notes that "the great majority of medieval Christian theologians held that God did not create a world necessarily and that, given that he did create, God was not bound by necessity to create just the world he in fact created. God's act of creation is, therefore, contingent in these two respects." A synchronic conception of contingency, MacDonald suggests, "is built into the foundations of the Christian understanding of God as free creator" (ibid., 171), so that "Scotus and his ilk can't have been the discoverers of these modal concepts" (ibid.) But this synchronic conception can be presupposed by the Christian understanding of God, without being fully articulated; one can interpret this as being characteristic of Ockham's position, and the source of his problems. One can likewise attribute to Scotus the merit of having fully articulated, and brought into the foreground, this presupposed conception.

35. This paper has benefitted very much from discussion on these matters with Calvin Normore and Martin Tweedale.

Aufredo Gonteri Brito *secundum*
Henry of Harclay on Divine Foreknowledge and Future Contingents*

Chris Schabel

The English secular theologian Henry of Harclay (ca. 1270-1317) played a significant academic role in both Paris and Oxford in the first two decades of the fourteenth century. From his days as a student of theology at Paris, we have a commentary on book one of the *Sentences* of Peter Lombard, presumably a *reportatio* of lectures from around 1300 which betray the influence of John Duns Scotus, then also at Paris. Surviving from a later period is a series of Questions, in which Harclay "shows more independence and maturity" as a theologian.[1]

While Harclay's Questions may be more mature, and have received the bulk of scholarly attention, it is his *Sentences* commentary that perhaps left the greater legacy in Paris. One finds citations of "Henry of England's" commentary in the giant *Scriptum* of Peter Auriol, OFM (? 1322), for example. Equally telling of Harclay's influence is the fact that Auriol's fellow Franciscan Aufredo Gonteri Brito incorporated almost the whole of Harclay's commentary into his own much larger work from the 1320's.[2] Indeed, the Scotist Gonteri's use of Harclay's text is an early and classic example of a scholar reading *secundum alium*, or "according to another."

Like much of Harclay's commentary, the questions presented below, i.e. distinctions 38 and 39 of book one, concerning divine foreknowledge and future contingents, were familiar to Auriol and used by Gonteri.[3] The problem of divine foreknowledge and future contingents is and has been for some time something of a "hot topic" in medieval intellectual history, and the present text has been the focus of some scrutiny. In his edition of Harclay's later "Questions on Divine Prescience and Predestination," Mark Henninger made a careful comparative study of the corresponding questions in Harclay's *Sentences* commentary, including distinctions 38 and 39.[4] Nearly half a century earlier, Hermann Schwamm had examined the questions on foreknowledge contained in Gonteri's commentary in ms. Vat. lat. 1113, although Schwamm knew neither Gonteri's identity nor his reliance on Harclay, nor even Harclay's commentary.[5] Nevertheless, Schwamm knew enough about the history of foreknowledge among Scotus's "ersten Anhängern" that he was able to differentiate between the new and the old. Hence most of Schwamm's discussion centered around Gonteri's additions to Harclay, and thus is a useful philosophical treatment of Gonteri on future contingents.[6]

Given the previous analysis of the present questions in the secondary literature, it will be more profitable here to concentrate on Gonteri's procedure in reading *secundum alium*. Although Damasus Trapp pointed to many instances of the phenomenon of reading *secundum alium*, especially from the decade between 1365 and 1375, not much has been done since then to follow up on his findings.[7] He found that, particularly among the Augustinians and Cistercians, one scholar would literally read the work of another while lecturing, sometimes without attribution. Reading *secundum alium* was not confined to this period, however. Besides the example of Gonteri, Dionysius de Burgo Sancti Sepulchri, OESA, and Bernardus Lombardus, OP, lecturing in 1317 and 1327 respectively, read much of Durandus of St. Pourçain verbatim in the context of future contingents, except where turning to oppose him.[8]

The practice raises a host of difficult questions about the nature of later-medieval teaching and writing procedures: What is reading *secundum alium* and what is not? The version of Henry of Langenstein's commentary *not* found in ms. Alençon 144 is simply James of Eltville's commentary verbatim, which Henry reread while visiting Abbot Eltville's Cistercian house at Eberbach.[9] But what of Pierre d'Ailly's book one, question eleven, which is merely a rearrangement, sometimes verbatim, sometimes paraphrased, of part of Gregory of Rimini's commentary?[10] What if a scholar read according to different theologians in varying contexts? James of Eltville himself, who Trapp says read *secundum* John Hiltalingen of Basel in many places, copied large sections of Alphonsus Vargas of Toledo, and to a lesser extent Gregory of Rimini, in the section on future contingents.[11] The most spectacular example of this cut and paste method is the *Sentences* commentary of what I call "the Vienna Group," a series of theologians at the University of Vienna in the first decades of the fifteenth century. One of them seems to have made a mix of Langenstein's Alençon commentary and Henry Totting of Oyta's mature commentary, and the others simply read it again, with the occasional emendation. Part of what they took from Langenstein concerning foreknowledge was in fact taken verbatim from Rimini![12]

Other questions come up: What is the connection between these writings and actual *Sentences* lectures, for which, supposedly, lengthy notes were not allowed? Did scholars read according to others covertly?

If so, was one ever found out? If openly, had he filled the requirements for his degree?

Few of these questions are answerable yet in the case of Aufredo Gonteri Brito. Gonteri probably heard Scotus lecture in Paris around 1303, and lectured on book one of the *Sentences* at the Franciscans' Barcelona convent in 1322, and again at Paris in 1325.[13] Most of Harclay's commentary has been absorbed into Gonteri's, but not all. For example, in distinction 39 Gonteri eliminates the sixth principal argument *ad oppositum* (ll. 39-60) and later its refutation (ll. 170-225).[14] These sections involve obligations, and either Gonteri thought the material too dense, too dated, or too tangential to copy. It may be more difficult to access the passages Gonteri adds. One would hesitate to assign any addition to Gonteri's originality, given his procedure otherwise. Indeed, Schwamm heard echoes of Hugh of Newcastle's commentary, and Charles Balic linked Gonteri with John of Reading, for example.[15] Gonteri's additions to Harclay's distinction 38, at ll. 36-83 and ll. 201-258, while not copied verbatim, I think can be safely attributed to the influence of Francis of Marchia, OFM, who read at Paris around 1319.

As Schwamm indicated, in the first addition Gonteri seems to be attacking Peter Auriol.[16] James Halverson reached the same conclusion about Gonteri's treatment of predestination, pointing to the very paragraph that Gonteri inserted into his Harclay material.[17] In fact in both cases Gonteri is paraphrasing his Franciscan confrère Marchia's attack on Auriol, based primarily on Marchia's idea that, contrary to Auriol's notion of contingency, there is some sort of determination *de inesse* in contingent causes before they act, via which God can know the future. All that was required to safeguard contingency, according to Marchia, was that the causes not be determinate *de possibili*. Thus for Marchia, a cause could be determined toward acting, but in a way that would not completely remove the possibility of not acting. Gonteri joins Michael of Massa, William of Rubione, Gregory of Rimini, and others in following Marchia here.[18]

In the second addition, Gonteri again follows Marchia, this time his refutations of earlier views on how God knows the future, refutations based in turn on Scotus, whom at the end of the addition Gonteri chooses to follow more closely than the more independent Marchia had.[19] In these questions, then, Gonteri shows himself to be a plagiarizer at worst, and completely derivative at best. How his *Sentences* commentary could stem from actual lectures, or whether he read Harclay openly or covertly, is as yet unknown. Gonteri's careful efforts to delete "inaccurate" internal reference found in the Harclay text (e.g. d. 38, l. 292) might be interpreted as covering up. Other sections of his work, however, appear to be less dependent on others.[20]

The Edition

The two surviving witnesses of Gonteri's commentary are helpful in the task of editing Harclay's own, much more important, *Sentences* commentary, itself surviving in only two manuscripts. I have worked entirely from microfilms of these four manuscripts:

C = Casale Monferrato, Seminario Vescovina, cod. B 2. 84 folios, 14th century: Harclay's *Sentences* commentary, book one. Present text, fol. 77ra-79va.

H = Biblioteca Apostolica Vaticana, Vat. lat. 13687. 190 folios, 14th century: Harlcay's *Sentences* commentary, book one (fol. 1-97vb). Present text, fol. 86vb-90rb.

A = Biblioteca Apostolica Vaticana, Vat. lat. 1113. 185 folios, 14th century: Gonteri's *Sentences* commentary, book one. Present text, fol. 168ra-169va (d. 38) and 169vb-170rb (d. 39).

W = Wroclaw, Biblioteka Uniwersytecka A 211 (formerly I.F. 184). 708 folios, early 15th century: Gonteri's *Sentences* commentary, books one and two. Present text, fol. 278ra-280rb (d. 38) and 280vb-281vb (d. 39).

It is probable that C stems from H. C has no less than 17 major omissions *per homoioteleuton* unique to it here, whereas H has no such singular omission. Occasionally H omits one word, or has a small error in mood or tense, but in each case the scribe of C would have expected or easily interpolated the missing word or the correct form. Examples are in d. 38, ll. 89, 196, and 281. Otherwise, C has scores of minor errors.

There are two difficulties with this theory. First, some early questions contained in H and C are not identical; hence we would have to conclude that C did not copy H in all sections. Second, at d. 39, l. 124, C and AW share an omission of eight words. That this is coincidental is plausible, because the text is often repetitious, and lends itself to frequent omissions *per homoioteleuton*. A similar instance occurs at d. 39, ll. 15-16, where C and W agree on an omission, but A has a slightly different omission.

As expected, CH differ from AW in scores of minor transpositions, omissions, and other small variants, as well as six shared omissions *per homoioteleuton*. A has nine singular major omissions, and W, although in a later hand, only three, although it has many small errors. Thus the manuscripts appear to be independent of each other. Apparently A records Gonteri's Barcelona lectures, and W is based on Paris, but there is little or no evidence of revision.[21] A was also corrected somewhat extensively. Unfortunately, it is usually not possible to tell whether AW's variants are to be attributed to the Harclay manuscript which Gonteri used, or to Gonteri's own changes.

There is a slight possibility that Gonteri used H in preparing his lectures. Occasionally a small error in CH is correct in AW, but in most cases the error is obvious (e.g. d. 38, l. 146), and the scribe could have corrected the text easily. More serious are three major omissions that C and H share, at d. 38, ll. 161-163, and at d. 39, ll. 19-20 and l. 133. In d. 38, l. 133, the text omitted is not necessary for sense, and may have been one of Gonteri's interpolations. A similar thing could be said for d. 39, ll. 19-20, where A also has an omission, and only W has the "correct" text. In the case of d. 38, ll. 161-163, however, it seems that CH have really made an error, and AW had to have had direct or indirect access to a previous exemplar. Perhaps some of the marginal additions of A, then, such as at d. 39, ll. 34, 36, 86 and 146, are actually corrections and not interpolations.

Thus here is the simplest possible stemma, by my findings:

```
        α
       / \
      H   β
     /   / \
    C   W   A
```

I have tried to follow H unless there was an obvious error, or unless AW had a minor addition of significant aid to the reader. For Gonteri's additions, except for major omissions, A and W are followed about equally. I include all major omissions in the apparatus criticus, and all singular variants for H, but none for A, C, and W, except where H has a variant, or in the cases when Gonteri has an omission or addition of material and only two manuscripts provide the text. I give all variants shared by A and W, as the Gonteri text. I do not include small orthographic variants, or the difference between 'ista' and 'illa.' H has some marginalia, which I record, except for the rephrasing of the question before both distinctions? I have stuck to the medieval 'e' instead of the classical 'ae' in the text. In the apparatus fontium, I have eliminated quotations. See Henninger's introduction and edition for corresponding passages in Harclay's later Questions, and for quotations from Aquinas and Scotus in Harclay.

University of Cyprus

Notes

* I wish to thank Katherine Tachau, William Courtenay, and the University of Cyprus Library for providing microfilm copies of manuscripts. I also thank Paul Spade, Jean-François Genest, William Duba, and Russell Friedman for helpful comments.

** Note that in the texts below, line numbers are provided in brackets: {}.

1. Mark Henninger, "Henry of Harclay's Questions on Divine Prescience and Predestination," *Franciscan Studies* 40 (1980), 167-168 (see Henninger's n. 2 for a bibliography on Harclay).

2. Cf. Charles Balic, "Adnotationes ad nonnullas quaestiones circa Ordinationem J. Duns Scoti" (*Iohannis Duns Scoti Opera Omnia* IV, Vatican City, 1956), 15*-28*.

3. For Auriol's citation of Harclay, see Petrus Aureoli, *Scriptum* I, d. 39, a. 3 (ed. Chris Schabel, "Peter Aureol on Divine Foreknowledge and Future Contingents: *Scriptum in Primum Librum Sententiarum,* distinctions 38-39," *Cahiers de l'Institut du Moyen Age Grec et Latin* 65 [1995], 185-6.803-11).

4. Cf. Henninger, "Henry of Harclay's," 183-93.

5. Cf. Hermann Schwamm, *Das göttliche Vorherwissen bei Duns Scotus und seinen ersten Anhängern* (Philosophie und Grenzwissenschaften, 5, 1-4) (Innsbruck, 1934), 187-206.

6. Of Schwamm's 20-page section on Vat. lat. 1113, only about four pages deal with material from Harclay, i.e. 191; 193-194; and 196. Gonteri adds several, mostly small, questions to dd. 38-9, and a *divisio textus*, all treated by Schwamm.

7. Cf. Damasus Trapp, "Augustinian Theology of the 14th Century: Notes on Editions, Marginalia, Opinions, and Book Lore," *Augustiniana* 6 (1956), 146-274, especially 250-5.

8. Cf. Chris Schabel, "The Quarrel with Aureol: Peter Aureol's Role in the Late Medieval Debate over Divine Foreknowledge and Future Contingents" (PhD Dissertation, University of Iowa 1994), 239-41.

9. Cf. MIchael Shank, *"Unless You Believe You Shall not Understand": Logic, University, and Society in Late Medieval Vienna* (Princeton: Princeton UP, 1982), 127, 215-7.

10. Cf. Schabel, "The Quarrel with Aureol," 330-335, especially the table of corresponding passages on 334.

11. For Eltville and Hiltelingen, see Trapp, "Augustinian Theology," 252. On Eltville's dependence on Vargas and Rimini, and his position on fuutre contingents, see Schabel, "The Quarrel with Aureol," 321-6.

12. On the "Vienna Group," i.e. Nicholas of Dinkelsbul, Peter of Pulkau, Arnold of Seehausen, and Johann Berwart of Villingen, and Johannes Wuel de Pruck, see Shank, *"Unless You Believe,"* 117-27. For a treatment of their (including Thomas Ebendorfer of Haselbach's) relationship with each other and with Langenstein, Oyta, and Rimini, in the context of future contingents, see Schabel, "The Quarrel with Aureol," 363-70.

13. Gonteri's dates are based on explicits cited in Katherine Tachau, *Vision and Certitude in the Age of Ockham: Optics, Epistemology, and the Foundations of Semantics 1250-1345* (Studien und Texte zur Geistesgeschichte des Mittelalters, 22) (Leiden: Brill, 1988), 318, n. 9. For a recent bibliography on Gonteri, see Stephen Dumont, "The Univocity of the Concept of Being in the Fourteenth Century: II. The *De Ente* of Peter Thomae," *Mediaeval Studies* 50 (1988), 194, n. 34, or Dumont's "The Scotist of Vat. lat. 869," *Archivum franciscanum historicum* 81 (1988), 281, nn. 2-3.

14. Balic ("Adnotationes," 16*) has already noted Gonteri's omission.

15. Cf. Schwamm, *Das göttliche Vorherwissen*, 235, and Balic, "Adnotationes," 21*.

16. Cf. Schwamm, *Das göttliche Vorherwissen*, 190.

17. Cf. James Halverson, "Peter Aureol and the Re-emergence of Predestinariam Pluralism in Latin Theology" (PhD Dissertation, University of Iowa 1993), 215-6.

18. Cf. Schabel, "The Quarrel with Aureol," 235-9; 254-5; and 272, n. 11.

19. Russell Friedman and I are presently working on the critical edition of book one of Marchia's *Sentences* commentary, which appears to survive in two distinct versions and one mixed. The citations to Marchia in the text will be to the *Commentarium*, represented by ms. Tortosa, Archivo capitular 162.

20. Cf. Tachau, *Vision and Certitude*, 322-7, and Dumont, "Univocity II", 193-6).

21. Cf. Dumont, "Univocity," 194, n. 34.

<HENRICUS DE HARCLAY ET AUFREDO GONTERI BRITO *IN I SENTENTIARUM*, DISTINCTIO 38>

Utrum Deus habeat determinatam scientiam futurorum contingentium?

[1] Quod non. Probatio: Primo *Posteriorum*, scientia est eorum que impossibile est aliter se habere. Sed contingens potest aliter se habere. Ergo eius non potest esse determinata scientia.

[2] Preterea, scientia determinata non est fallibilis. Sed scientia infallibilis non est contingens. Ergo a primo, etc. Probatio medie, quod scientia infallibilis non est contingens, quia tunc esset fallibilis. Probatio: Scit Deus me sedere cras. Si ergo possum non sedere cras, ponatur in esse. Non accidit impossibile, 'Non sedebo cras'. Ergo Deus scit me sedere et non sedebo. Ergo fallitur.

{10}[3] Preterea, sequitur, 'Deus scit aliter quam est in re; ergo fallitur'. Ergo eadem ratione sequetur, 'Deus scit aliter quam potest esse; ergo potest falli'. Probatio quod sequatur, quia sicut ex duabus de inesse sequi-tur conclusio de inesse, ita ex una de possibili et altera de inesse sequitur conclusio de possibili. Ergo sequitur, 'Deus scit hoc esse; et hoc potest non esse; ergo potest falli'. Sed maior et minor est vera. Ergo conclusio, scilicet, 'Deus potest falli'.

[4] Preterea, scientia determinata limitata est ad unum. Sed scientia Dei non potest limitari. Ergo non ad unum determinari. Probatio maioris, nam potentia que est determinata ad unum effectum est potentia limitata. Ergo eodem modo de scientia.

[5] Preterea, primo *Peryarmenias*, in futuris contingentibus non est veritas determinata. Et probat, quia secundum hoc non oporteret consiliari neque negotiari. Sed si scientia nostra facit determinationem in scito, vel arguit multo magis scientia Dei que est certiorum. Ergo cum contingenti repugnet certitudo, ei repugnat cognosci determinate a Deo.

[6] Preterea, omne scitum a Deo de necessitate est verum. Contingens futurum a Deo est scitum. Ergo de necessitate est verum. Maior est vera, quia scitum a nobis de necessitate est verum. Probatio consequentie, quia ex maiori de necessario et minori de inesse simpliciter, sequitur conclusio de necessario, primo *Priorum*. Sed hec minor est de inesse simpliciter. Probatio: Illa propositio de inesse est de inesse simpliciter in qua ponitur illud quod non potest separari a subiecto, ut cum dicitur, 'Corvus est niger', 'Ignis est calidus'. Sed ab hoc subiecto, 'futurum contingens', non potest separari hoc predicatum, 'esse scitum a Deo'. Ergo est de in-esse simpliciter.

Oppositum, Ecclesiastici 23, et ponitur in littera, primo capitulo "Domino Deo antequam crearentur omnia sunt agnita."

Preterea, 15 *De trinitate*, capitulo 13, et ponitur in littera, capitulo primo, dicit Augustinus: "Non aliter scivit creata quam creanda." Sed creatorum habet determinatam scientiam et infallibilem. Ergo et creandorum.

<Additio prima Aufredi>

In ista questione primo videndum est quomodo et quot modis est contingentia in rebus? Ad cuius evidentiam sciendum est quod triplex est contingens, scilicet in pluribus, et ut raro, et ad utrumlibet. Contingens ut in pluribus intenditur in causis naturalibus impedibilibus, et contingens raro, sive a casu, ei opponitur et evenit per accidens. Et in talibus est contingentia extrinseca et per accidens, quia cause naturales, omni impedimento excluso, necessario causant. Et ideo potens cognoscere omnem concursum causarum impedimentium, potest determinate cognoscere effectum contingentem. Et quia Deus determinate cognoscit omnes causas naturales que in se sunt prius determinate quam causent, potest cognoscere certitudinaliter effectus tales naturales. Unde et respectu eius nullus effectus videtur casualiter evenire.

Sed in causa contingenter agente ad utrumlibet et libere, qua potest effectus poni vel non poni, attenditur contingentia per se et intrinseca, secundum quod aliquis per liberum arbitrium est principium suorum actuum virtuosorum vel vitiosorum. Et causa prima huius contingentie est divina voluntas, que est causa prima omnium motionum, secundum Augustinum, 3 *De trinitate*, que si necessario ageret, nihil esset contingens in rebus.

Secundo videndum est utrum effectus contingens sit prius determinatus in causa contingente quam ponatur in esse, ac per hoc sit cognoscibilis? Et dicunt aliqui quod non, quia effectus non potest magis determinate cognosci per causam quam contineatur in causa. Sed effectus continetur indeterminate et sub indifferentia ad alteram partem in causa contingente -- per hoc distinguitur a causa necessaria. Ergo etc.

Preterea, illa determinatio cause contingentis ad producendum effectum contingentem aut est in eius potestate aut non. Si non, ergo nec sua actio nec effectus contingens. Si sic, quero per quid, quia non per actionem cause, quia eam precedit, et posterius non est causa prioris. Si per rationem formalem cause, ergo causa ex se est determinata, et ita non est contingens.

Preterea, effectus qui non cognoscitur positis omnibus suis causis, non cognoscitur determinate per eas. Effectus contingens est huiusmodi, quia de ratione eius est quod possit poni et non poni, positis omnibus suis causis.

Sed istud non est verum, quia omnis causa que est ratio essendi effectum determinate quantum ad unam partem contradictionis et non aliam potest esse ratio determinate cognoscendi effectum illum respectu intellectus perfecte cognoscentis eam, quia sicut continet in entitate ita in cognoscibilitate. Sed causa contingens est huiusmodi, quia quando effectus contingens ponitur in esse, ponitur secundum unam partem contradictionis determinate. Sed tunc ponitur a causa contingente.

Confirmatur, quia Deus habet notitiam futurorum contingentium determinatam antequam eveniant, sicut quando eveniunt, quia, "Non aliter cognovit fienda quam facta," secundum Augustinum, 5 *Super Genesim*. Nec habet de eis coniecturalem notitiam, nec sub disiunctione, quia hoc ponit imperfectionem in eius notitia. Nec cognoscit ea per rationes proprias eorum, tum quia in se non existunt, et <tum> quia vilescet eius intellectus. Ergo cognoscit ea per causam contingentem, non necessariam.

Preterea, frustra esset consilium in agente a proposito de agendis contingenter, nisi prius esset causa contingens determinata quam causet contingenter. Licet ergo absolute causa contingens sit indifferens ad posse agere et non posse agere, et ad agere et non agere, prius tamen determinatur libere et contingenter quam agat. Et dico quod ex se determinat se ad {80} agendum, et dum determinat potest se non determinare in sensu diviso.

Et per hoc patet responsio ad rationes opinionis.

Tertio videndum est per quid et quomodo intellectus divinus intelligat futura contingentia determinate vera? <*Finis additionis primae Aufredi.*> Dicitur uno modo, et est opinio Thome, parte prima, questione 14, 13 articulo, et in *Scripto* questione 225, quod contingens potest dupliciter considerari, vel in se, vel in comparatione ad suam causam. Primo modo non attenditur rei contingentia, scilicet secundum esse eius actuale, quia secundum illud esse est determinatum et necessarium, quia "omne quod est, quando est, etc."

Si autem consideretur contingens in comparatione ad suam causam, hoc potest esse dupliciter, vel ad causam propinquam, vel ad remotam. Si in comparatione eius ad causam {90} remotam, non habet contingentiam nec necessitatem, sed tantum in comparatione ad causam propinquam, nam causa remota existente necessaria, potest nihilominus effectus esse contingens, sicut ponit exemplum: Nam motus solis est causa remota floritionis arboris, et ideo potest contingenter florere arbor, etsi sol moveatur in eternum. Sed causa propinqua est virtus generativa plante, qua posita, necessario ponitur floritio, et ideo respectu illius non est contingens. Modo ad propositum, scientia Dei est causa remota omnium entium. Igitur etsi ipsa sit necessaria, nihil prohibet effectum esse contingentem. Et hoc modo non habet Deus determinatam scientiam futurorum contingentium, scilicet in comparatione ad causas contingentes a quibus contingenter dependent.

Sed quia Deus non tantum cognoscit res ut habent esse in suis causis, sed ut habent {100} esse in se presentialiter, et ut sic sunt determinate et non contingenter, ideo determinatam scientiam habet Deus de illis ut sint. Ratio ad hoc ponitur talis, nam omnia futura et preterita sunt Deo presentia, sicut sunt nobis presentia cum actualiter existunt. Ideo, sicut nos habemus determinatam cognitionem rerum cum sunt in actu, ita et Ipse semper, quia semper sunt sibi presentia.

Exemplum ponit istius et rationem. Ratio est ista, quia licet sit successio in tempore, tamen unum nunc eternitatis est simul cum toto tempore. Et hoc probat per Boethium in fine libri *De consolatione*, ubi dicit quod quia omnis cognitio est per modum cognoscentis, cum ergo Deus sit eternus, oportet quod eius cognitio sit eterna.

Ponit aliud exemplum in *Scripto*: Esto quod quinque homines in quinque horis successive, vel unus homo in quinque horis videat quinque contingentia {110} successive. Certum est quod videt illa diversis intuitionibus et presentialiter, successive tamen. Esto ergo quod <si> omnes iste visiones essent unus actus cognitionis, simul videret presentialiter omnes. Aliud exemplum ponitur de existente in turre, etc.

Ratio istorum est ista, quia si esset tempus totum simul, non repugnaret uni nunc eternitatis esse cum toto tempore. Sed nihil deperit uni nunc eternitatis de immensitate sua, ratione cuius convenit sibi esse simul cum omni parte temporis per hoc quod tempus est successivum. Verbi gratia, baculus fixus in aqua vel fluvio est simul cum toto fluvio successive. Quod autem non est simul cum toto fluvio, hoc est quia non est immensus. Sed nunc eternitas est <immensum>. Ideo etc. Ideo, quia Deus presens est in eternitate cuilibet parti {120} temporis, ideo dicendum, sicut determinate novit rem cum est in actu, ita determinate novit eam quando non est in actu.

Contra ista. Primo contra hoc, quia dicit quod res, ut consideratur sub actuali existentia, non est contingens. Ergo rei non potest competere contingentia dum habet actualem existentiam. Ergo si res haberet actualem existentiam eternam, non posset esse contingens, quia tota contingentia rei per ipsum est per comparationem ad causam que potest contingenter causare, et dum res est non contingenter causatur. Ergo oportet causam contingentem precedere effectum causatum, et per consequens si res esset eternaliter existens, nulla posset esse contingentia in re. Ergo ex opposito, si res quelibet causata necessario est contingens et omne aliud a Deo, impossibile est quod aliquod ens aliud a Deo sit eternum. {130} Hoc est contra ipsum qui dicit mundum potuisse fuisse ab eterno.

Preterea, illud quod non existit, illud est non-ens, et antequam crearetur nihil fuit. Sed contingens non est non-ens nec nihil. Ergo si sit, contingens manifestum erit dum existit. Probatio minoris: Certum est quod non omne ens est necesse esse. Ergo aliquod ens est contingens. Sed non est nisi dum presentialiter existit, aliter semper fuisset ens et numquam crearetur.

Dices, 'Immo, est contingens hec, "Sedens ambulat", et hec est vera de contingenti, "Possibile est sedentem ambulare", et tamen ambulans non existit'.

Respondeo. Non est ad propositum, nam verum est quod contingentia, ut facit propositionem modalem, compatitur non-existentiam extremorum illius propositionis, sicut {140} dicit Aristoteles, primo *Priorum*, "B potest esse A', quia talis contingentia non significat nisi non-repugnantiam extremorum respectu esse, vel pro uno tempore pro sensu composito, vel pro diversis temporibus secundum sensum divisum.

Sed contingentia ut est differentia alicuius entis incomplexi presupponit actualem entitatem in illo cuius est conditio.

Preterea, exemplum suum de floritione non valet. Probatio: Ista causa, si sit causa, non est contingens vel remota qua posita et stante impossibile est effectum non esse. Sed posita scientia Dei respectu A esse, et stante illa, impossibile est simpliciter A non esse. Ergo si scientia Dei sit causa, non erit causa remota, ut se habet motus solis respectu floritionis arboris. Et ipsemet obicit hoc modo in *Scripto*. Et respondet modo quo dictum est supra. Sed saltem illa ratio de floritione non valet, quia illam non defendit in sua responsione.

Contra illud quod dicit, quod Deus est presens omni parti temporis, contra: Quod nihil est in se nulli est presens. Sed antequam res sit, nihil est in se. Ergo nulli est presens.

Quando ipse dicit quod propter suam immensitatem est presens cuilibet parti temporis, dico quod sua immensitas et illimitatio non se extendit ad essendum simul cum eo quod nihil est, sed tantum ad essendum cum omni eo quod est, ita quod impossibile sit aliquid esse nisi Deus sit cum illo. Probatio istius: Deus non est in illo vel presens illi quod nihil participat sue entitatis vel sue bonitatis, quia suum esse cum creaturis non est nisi bonitatem suam et entitatem suam eis communicare vel influere et illa participare. Sed non-ens vel nihil-ens non participat aliquid entitatis, quia eo ipso esset ens. Ergo cum illo non est Deus presens.

Quando dicis, 'Si essent partes temporis simul, esset quidem unum nunc eternitatis presens illis omnibus', concedo.

Tu dicis, 'Nihil deperit immensitati per hoc quod partes temporis sibi succedunt'. Dico quod verum est, tamen in tantum deperit huic relationi simultatis, quia alterum extremum non est. Verbi gratia, Deus per suam infinitatem est dominus totius creature, et cum desinit esse creatura, nihil deperit Deo. Tamen secundum Augustinum, 5 *De trinitate*, non amplius esset dominus quia non haberet creaturam servam. Ideo deperiret alterum extremum relationis.

Exemplum de stante in turre et alia non valent, quia tales potest videre transeuntes. Quid mirum, quia illi sunt visibilies et existentes. Sed res preterita vel futura in se est nonens et nihil, et per consequens non visibilis.

Dico ergo quod Deus habet determinatam scientiam futurorum contingentium et infallibilem. Dico secundo quod non habet necessariam scientiam illorum. Probatio primi satis facilis, nam rem existentem novit determinate -- ita possem ego. Sed secundum Augustinum, 15 *De trinitate*, capitulo 13, "Non aliter novit creata quam creanda," et Apostolum, ad Romanos 4, "Vocat ea que non sunt tamquam ea que sunt." Ergo determinate novit antequam fierent sicut postea.

Preterea, 6 *De trinitate*, ultimo, "Accendant vel succedant tempora; nihil accedit vel succedit scientie Dei." Ergo uniformiter cognoscit.

{180} Secundum probatur, quod eius scientia non est necessaria respectu futurorum contingentium. Probatio istius: Illud quod eo solo est scitum quia est volitum, si non sit necessario volitum, non est necessario scitum -- istud est planum. Sed contingens ad utrumlibet non est scitum nisi quia volitum. Ergo etc. Probatio minoris: Notitia propositionis contingentis non est notitia simplicis quidditatis extremorum. Tunc enim talis propositio non esset contingens, sed esset simpliciter necessaria et per se nota. Sed oblatis intellectui duabus quidditatibus quarum coniunctio in esse non est ex natura illarum, verbi gratia, Sortes ex parte una et beatitudo ex parte alia, cognito quid sit Sortes et quid beatitudo, non propter hoc cognoscitur ista, 'Sortes est beatus'. Ergo quod ista cognoscatur oportet eius veritatem determinari per voluntatem Dei, quia ex natura terminorum non {190} potest haberi veritas eius. Ergo pro tanto cognoscitur illa, 'Petrus erit beatus', pro quanto voluntas divina determinatur ad coniungendum ista extrema, scilicet Petrum et beatitudinem. Sed voluntas divina non necessario hoc determinat. Ergo nec hoc necessario novit.

Secus autem est de propositionibus simpliciter necessariis, nam coniunctio illorum extremorum non determinatur a voluntate divina, sed necessitate nature vel naturali, cuius oppositum includit contradictionem. Coniunguntur ista duo, 'diameter' et 'incommensurabilitas coste', quia oblatis istis duabus, non expectatur determinatio voluntatis divine respectu coniunctionis illorum, quia non possunt non coniungi. Et ideo cognitis illis extremis eo ipso necessario cognoscit complexum compositum ex ipsis, etiam si {200} per impossibile non esset volens.

<Additio secunda Aufredi>

Dico ergo quod non cognoscit per rationem divine essentie vel per rationem idee precise, quia essentia divina et idea, quomodocumque ponatur, representat necessario uniformiter determinate, prius omni actu voluntatis. Sic autem non possunt representari futura contingentia, nec sunt determinate vera ad unam partem, quia qua ratione secundum unam partem representarentur, eadem ratione secundum aliam, vel secundum ambas simul, quod est impossibile.

Confirmatur, quia per illud quod representat terminos simplices alicuius complexi non representatur illud complexum, nisi ex natura terminorum cognoscatur. Sed essentia divina vel idea representat quidditates simplices ex natura terminorum, et ex natura {210} terminorum non possunt futura contingentia determinate cognosci esse vera, quia tunc essent necessaria et evidentia ex terminis. Ergo etc.

Dico ergo quod non cognoscuntur per hoc quod intuitus divinus eternus, ratione sue immensitatis, fertur super ea antequam existant, sicut si existerent, sicut dicunt quidam. Tum quia quod in se non existit nulli coexistit. Futura contingentia in se non existunt. Ergo etc. Tum quia futura contingentia Deus non cognoscit determinate per rationem eorum tamquam rationem et medium cognoscendi, sed per aliquid in se existens per quod determinatur ad existere in futuro. Aliter vilesceret eius intellectus.

Dico ergo, secundum mentem Doctoris Subtilis, quod licet intellectus divinus per rationem essentie divine cognoscat terminos simplices complexionum contingentium, et ipsas {220} complexiones ut neutras, ipsas tamen ut determinate veras non cognoscit nisi prima determinatione sue voluntatis libere, qua ab eterno determinate voluit futura in re evenire. Sed hoc non est ex imperfectione divine essentie et divini intellectus, sed ex imperfectione futurorum contingentium, que non sunt nata esse determinate vera nisi quia volita fore, sicut nec respectus rationis natus est determinate cognosci per rationem essentie divine nisi presupposito actu collativo intellectus. Et ita determinatio divine voluntatis est causa propinqua qua sunt determinate cognoscibilia ab intellectu divino, et per actum intelligendi formaliter representantur.

Et si arguatur, sicut quidam arguunt contra hoc, quod tunc voluntas divina, determinans intellectum divinum ad determinate cognoscendum futura contingentia, esset {230} nobilior intellectu sicut in nobis, dico quod nullum est inconveniens voluntatem divinam secundum rationem esse nobiliorem intellectu, sicut in nobis realiter actus voluntatis est nobilior actu intellectus, cum sint diversarum specierum. Et species in perfectione necessario sunt inequales, quia qualem ordinem habent aliqua in prioritate et perfectione ubi distinguuntur realiter, talem habent ubi distinguuntur secundum rationem. Sicut essentia divina dicitur nobilior et perfectior simpliciter quam proprietas personalis, unam tamen perfectionem simpliciter realiter dicunt.

Et hoc dicit Magister in littera quod scientia Dei prout dicit precise notitiam intellectus est causa sine qua non rerum. Sed divinum beneplacitum et divina voluntas est causa prima et principalis omnium, secundum Augustinum, tertio *De trinitate*. Dico etiam {240} quod determinationem divine voluntatis preexigi ad cognoscendum determinate contingentia futura ut vera non ponit imperfectionem in divina essentia vel intellectu, sed tantum in futuris contingentibus.

Item arguitur, quia intellectus divinus, intelligens futura contingentia determinate vera, aut intelligit ea per essentiam divinam aut per determinationem divine voluntatis. Si per determinationem divine voluntatis, contra: Quia intellectus qui eodem actu intelligit necessaria et contingentia, eodem modo utraque intelligit. Intellectus divinus est huiusmodi, quia uno actu simplici omnia intelligit. Si dicatur quod intelligit ea per essentiam, contra: Quia intellectus perfecte non intelligit aliquid nisi eius primam causam intelligat. Sed essentia secundum quod essentia, et non determinatio divine voluntatis, est {250} prima causa veritatis et entitatis futurorum contingentium. Ergo etc.

Dico quod causa representativa activa veritatis determinate futurorum contingentium in intellectu divino est determinatio divine voluntatis, quia non sunt nata sciri esse determinate vera nisi quia sunt volita. Sed ratio formalis qua representantur ut obiecta secundaria est actus divini intellectus invariabilis et unus formaliter ex unitate obiecti sui primi, quod est essentia divina. Exemplum de albente et causante albedinem, in qua consequitur similitudo ad aliam albedinem. Cum variatione enim et multiplicatione obiectorum secundariorum stat invariatio et unitas actus respectu obiecti primi.

Per hoc patet faciliter ad rationes istius. <*Finis additionis secundae Aufredi.*>

[~1] Ad primum argumentum, dico quod scientia tripliciter accipitur. Aliquando {260} scientia extenditur ad fidem, quomodo accipit Augustinus, 15 *De trinitate* 12: "Absit ut scire nos negemus que testimonio aliorum didicimus." Aliquando accipitur specialiter pro cognitione certa rei necessarie, et sic accipit Aristoteles, primo *Posteriorum*. Et hec cognitio dupliciter, nam quedam est rerum temporalium, et hec proprie dicitur 'scientia'. Quedam vero est cognitio certa rerum spiritualium et eternarum, et hec vocatur 'sapientia'. Istam distinctionem ponit Augustinus, 12 *De trinitate*, ultimo capitulo. Sed non est necesse quod cognitio certa sit obiecti necessarii, nam Sorte currente, potest homo certitudinaliter scire eum currere, cum Sortem currere non est necessarium nisi sub conditione, scilicet dum currit. Et similiter concedo quod dum scitur a Deo, necessario erit in sensu composito.

[~2] Ad secundum, dico quod non est scientia eius fallibilis. Cum dicis, 'A potest non {270} esse, et ponatur in esse', respondeo quod si illud ponatur in esse, propter eius positionem non accidet impossibile, sed quia resultat incompossibilitas ex positione illius in esse simul cum alio in esse, quia scilicet ponitur ista, 'Deus scit hoc esse', et cum hoc ponitur simul non esse, 'Hoc non est', quia ista contradicunt, ideo, si unum ponatur in esse, oportet alterum removeri ab esse, sicut patet in multis exemplis.

[~3] Ad tertium, dico quod non sequitur, 'Estimat aliter quam est; ergo fallitur; ergo si estimat aliter quam potest esse, quod fallatur'. Responsio est ista: Nam posse falli non est posse scientiam mutari vel variari, sed posse falli est quod ista duo sint simul vera, 'Hoc credo' et 'Hoc non est'. Verbi gratia, ego scio me sedere nunc. Modo si non sedeam, scio etiam me non sedere. Ergo si mutaret de non sedente in sedentem, mutaretur etiam scientia mea, et {280} tamen numquam fallerer. Immo ex hoc sequitur quod non fallor, quia extimatio sequitur mutationem rei. Ideo est opinio vera et non fallit, quia in "eo quod res est vel non est dicitur oratio vera vel falsa." Et ideo non sequitur quod posset falli.

Ad probationem, dico quod ex altera de inesse simpliciter et altera de possibili, sequitur conclusio de possibili. Sed hec maior, 'Deus scit hoc esse', non est de inesse simpliciter, quia propositio de inesse simpliciter non differt realiter ab illa de necessario. Sed hec non est necessaria, sicut dictum est, et dicetur in fine.

[~4] Ad quartum, dico quod potentia que non potest nisi in unum effectum est limitata. Ita dico quod scientia que non potest nisi in unum scitum, si oppositum illius sciti est possibile, et scientia non posset in illud, esset limitata. Sed sic non est in proposito, nam {290} sicut possibile est A esse, tam possibile est A sciri a Deo esse, et sicut possibile est A non esse, ita possibile est Deum scire non esse et non scire esse, sicut dicetur infra. Tamen non poterit mutari, ut dicetur sequenti questione, distinctione 39. Sed scientiam esse in actu unius ita quod non alterius, scilicet contradictorii, non est illam esse limitatam, quia nulla illimitatio potest facere scientiam esse duorum contradictoriorum, sed scientiam esse unius ita quod non potest esse alterius obiecti non contradictorii. Sed disparatu hoc, arguit ipsam esse limitatam. Sic non est de scientia Dei.

[~5] Ad quintum, dico quod in futuris contingentibus non est veritas determinata a parte rei in se nec a parte suarum causarum, id est non est necessaria. Tamen eventus est determinatus apud Deum, etsi non sit <necessarius>. Sed tunc oritur difficillima ratio mihi, {300} nam esto quod non necessario scit A fore, sed contingenter tamen determinate scit A fore sicut quando A est. Et quando est, non necessario est. Ideo numquam scitur necessario. Tamen cum hoc ponitur in esse, non est tunc circa hoc negotiandum vel consiliandum. Sed antequam ponatur in esse eodem modo, quia immutabiliter scitur a Deo A esse, etsi non necessario, non est consiliandum de hoc utrum sit agendum vel non.

Dicetur, 'Verum est, si sciretur que pars esset determinata a Deo. Sed hoc non potest sciri. Ideo vel a Deo'. Sed contra: Quantumcumque sit mihi ignotum utrum A eveniet, cum eius oppositum numquam evenit, si sit determinatum a Deo vel eius opposita, ergo numquam per meum consilium impedietur vel promovebitur negotium.

Circa solutionem istius laboravit Augustinus, *De predestinatione sanctorum.* Anselmus, *De concordia predestinationis vel prescientie et liberi arbitrii.* Et ratio illorum stat in hoc quod sicut Deus scit me peccaturum, ita scit me peccaturum nonnisi per liberum arbitrium et voluntatem. Et ideo sequitur magis oppositum, quia si illud quod scitur a Deo futurum erit, cum ipse scit me peccaturum non necessario, sed libere, sequitur quod non necessario peccabo, sed libere. Eodem modo, sicut scivit hunc effectum contingentem productum a me, ita scivit quod non-nisi mediante deliberatione et consilio producam. Et ideo oportet consiliari et negotiari. Unde Anselmus in libro predicto distinguit necessitatem in antecedentem, que efficit, et necessitatem sequentem, que nihil efficit.

[~6] Ad ultimum, dico quod illa minor, 'Futurum contingens est scitum a Deo', non est de inesse simpliciter, sed de inesse ut nunc. Et ex maiori de necessario et minori de inesse ut nunc, non sequitur conclusio de necessario.

Doctor tamen Thomas, *Scripto* questione 225, et *Summa* in parte <prima>, questione 14, articulo 13, ponit multas responsiones quas omnes excludit, et ultimo stat in hoc quod ista est simpliciter necessaria, 'Hoc est prescitum a Deo'. Et innuit probationi ad hoc quia dicit quod est necessaria propter immobilitatem actus sciendi, et prima contingentia repugnat Deo, et per consequens scientie divine que est eadem cum sua essentia. Ideo est hec necessaria, 'Deus scit a fore'.

Si arguatur, '"Deus scit A fore; ergo A erit"; consequens est contingens; ergo antecedens', respondetur quod consequens ut sequitur ex antecedente est necessarium et non contingens, et hoc sic quia "in antecedente ponitur aliquid pertinens ad actum anime," cum Deus scit A fore. Ideo in consequente sic accipiendum est. Cadit autem res con-tingens sub consideratione Dei nisi dum est in actu. Hec sunt verba eius. Et res dum est in actu habet certitudinem et non contingentiam, ut dictum est supra. Ideo hoc consequens: 'A erit', ut sequitur ex hoc antecedente, 'Deus scit A fore', est sic intelligendum, 'A erit ut prescitur a Deo', hoc est secundum esse actualis existentie, et sic est necessarium. Et allegat ad hoc, "Omne quod est, quando est, etc."

Contra illud, primo respondeo ad rationem qua probat propositum. Dico quod non sequitur, 'Actus intelligendi Dei ut cadit super obiectum est immobilis et immutabilis; ergo est necessarius', quia actus volendi est equaliter idem cum Deo sicut actus sciendi, et actus volendi Petrum esse beatum est immutabilis, nam non potest Deus habere volitionem novam, et tamen non est necessarius. Probatio: Tunc nihil esset contingens. Probatio: Si enim Deus necessario vult hoc esse, et necessario potest, manifestum ergo necessario facit. Ergo hoc est necessario factum et non potest non fieri. Ergo non contingens. Consequens est falsum.

Et cum dicis, 'Deo repugnat contingentia', illud non valet, nam eo ipso quod scientia Dei est in se, est ens necessarium. Et obiectum complexum est contingens et non necessarium eo ipso scientia Dei ex parte sui contingenter respicit illud obiectum, quia si ex parte sua vel sui necessario caderet super illud, coexigeret illud. Sed necessarium non potest coexigere contingens.

Preterea, contra hoc quod dicit quod res dum est in effectu non habet contingentiam, illud est contra eum, quia tunc si res semper esset in actu, numquam esset contingens. Et tunc {350} arguo: Creature repugnat esse non contingens. Ergo ei repugnat eternitas.

Preterea, quero intellectum huius propositionis, cum dicit expresse in *Scripto*, "Res non ponitur subiacere scientie divine, nisi dum est in actu." Aut intelligis ita quod cessante re ab esse actuali cessat divina cognitio, et illud est falsum. Aut intelligis hoc modo quod Deus cognoscit rem dum est in actu, ita quod sciat rem non posse non esse dum est in actu. Et tunc redit argumentum precedens, quia tunc numquam esset res contingens, si fuisset ab eterno, quia numquam precessit potentia suum esse duratione, quod tamen oporteret ad hoc quod esset contingens.

Preterea, quod ipse dicit, 'Consequens esse necessarium non sequitur ex antecedente', illud est falsum. Probatio: A contingenter et non necessario erit. Ergo Deus scit A {360} contingenter et non necessario esse. Ergo hoc consequens ut cadit sub scientia Dei ut sic est contingens. Et ita dicit Augustinus et Anselmus.

Preterea, sophistice nimis respondet ad hoc argumentum, nam si consequens est necessarium ut sequitur ex antecedente, et simpliciter sequitur ex antecedente, et antecedens est simpliciter verum, ergo consequens erit simpliciter verum et sine conditione, quia ita est antecedens.

Confirmatur: Nulla propositio est in mundo ita necessaria, nec ista, 'Homo est animal', quin aliquam conditionem requirat ad sui veritatem. Quia tamen ista conditio est simpliciter necessaria, ideo simpliciter hoc est necessarium et sine conditione contingente. Et per te hoc, 'Deus scit A fore', est simpliciter necessarium. Ergo consequens simpliciter {370} necessarium.

Ultimo contra illud: Si hec sit necessaria, 'Deus scit A esse', eius contradictoria est impossibilis, 'Deus non scit A esse'. Consequens falsum, quia hec est possibilis. Probatio: Quia potest non scire A esse. Sequitur enim, 'A potest non esse; ergo Deus potest scire A non esse'. Sed cum Deus non potest scire contradictoria, cum scit A non esse, sequitur quod non scit A esse. Probatio istius: Quia contradictoria relata ad idem instans -- aliter non sunt contradictoria -- non sunt intelligibilia, nec scibile, nec cognoscibile, ergo impossibile est ista esse simul vera, 'Deus scit A esse', et 'Deus scit A non esse'. Ergo oppositum potest esse, scilicet, 'Non sit A', immo oppositum sequitur, quia cum quo non stat unum oppositorum, ad idem sequitur reliquum.

Apparatus Fontium

1-2 Cf. Aristoteles, *Analytica Posteriora* I, c 2: 71b15-16 (AL 4.1, 7.10-11); I, c. 4: 73a21-22 (AL 4.1, 12.8-9).
19-20 Cf. Aristoteles, *De interpretatione* I, c. 9: 18a29-19b4 (AL 2.1, 13.12-18.4).
25-26 Cf. Aristoteles, *Analytica Priora* I, c. 9: 30a15-23 (AL 3.1, 21.3-11).
31-32 Eccli. 23.29. Cf. Petrus Lombardus, *Sententiae in IV libris distinctae* I, d. 38, c. 1 (v.1, ed. Grottaferrata 1971, 276.15-16).
33-34 Augustinus, *De trinitate* XV, c. 13, n. 22 (CCL 50a, 495). Cf. Lombardus, *I Sent.*, d. 38, c. 1 (276.13).
37-50 Cf. Franciscus de Marchia, *Commentarium in primum Sententiarum* I, d. 36, a. 1 (ms. Tortosa, Archivo Capitular 162, fol. 47rb-va). Cf. Durandus de Sancto Porciano, *In Petri Lombardi Sententias theologicas commentariorum libri IIII*, I, d. 38, q. 3 (ed. Venice 1571, fol. 104rb).
50 Cf. Augustinus, *De trinitate* III, c. 4, n. 9 (CCL 50, 136.23-24).
51-80 Cf. Marchia, *I Sent.*, d. 35 (ed. Schabel, "The Quarrel with Aureol," 715-719.33-235), et d. 36, a. 2 (ms. Tortosa, fol. 47va-48ra).
52 Scil. Petrus Aureoli. Cf. Aureoli, *Scriptum*, d. 38, a. 1 (ed. cit., 96-98.197-226).
71-72 Augustinus, *De civitate Dei* X, c. 12 (CCL 48, 287.30-31).
83-121 Cf. Thomas Aquinas, *Scriptum super libros Sententiarum* I, d. 38, q. 1, a. 5 (vol. 1, ed. Mandonnet, Paris 1929, 909-911), et *Sum-ma theologica* I, q. 14, a. 13 (ed. Paulinae, Rome 1918, 83a-b).
87 Aristoteles, *De interpretatione* I, c. 9: 19a23-24 (AL 2.1, 17.2-3).
117-118 Exemplum est Scoti. Cf. Iohannes Duns Scotus, *Ordinatio* I, d. 38 pars 2 et d. 39 qq. 1-5 (v. 6, ed. Balic, Vatican City 1963, 408.4-10).
106-108 Cf. Boethius, *De consolatione Philosophiae* V, prosa 6 (CCL 94, 102-103.68-80). Cf. Aquinas, *I Sent.*, d. 38, q. 1, a. 5 (ed. cit., 911).
109-112 Cf. Aquinas, *I Sent.*, d. 38, q. 1, a. 5 (ed. cit., 911).
175 Augustinus, *De trinitate* XV, c. 13, n. 22 (CCL 50a, 495).
176 Rom. 4.17.
178-179 Augustinus, *De trinitate* VI, c. 10, n. 11 (CCL 50, 240.25-26).
201-257 Cf. Marchia, *I Sent.*, d. 36, a. 4 (ms. Tortosa, fol. 48va-b).
212-213 Cf. Ricardus de Mediavilla, *Super quatuor libros Sententiarum* I, d. 39, q. 1 (ed. Brixiae 1591, 344b-345a)..
218-227 Cf. Scotus, *Ordinatio* I, d. 38 pars 2 et d. 39 qq. 1-5 (ed. cit., 428.5-429.5).

228 Cf. Aureoli, *Reportatio in primum librum Sententiarum*, d. 38, q. 2 (ms. Berlin, Staatsbibl. Theol. Fol. 536, fol. 92vb-93ra).

237-238 Cf. Lombardus, *I Sent.*, d. 38, c. 1 (ed. cit., 277.23-25).

238-239 Cf. Augustinus, *De trinitate* III, c. 4, n. 9 (CCL 50, 136.23-24).

260-261 Augustinus, *De trinitate* XV, c. 12, n. 21 (CCL 50a, 493.75-76).

261-262 Cf. Aristoteles, *Analytica posteriora* I, c. 2: 71b17-19 (AL 4.1, 7.12-14).

263-265 Cf. Augustinus, *De trinitate* XII, c. 15 (CCL 50a, 379.41-45).

281-282 Aristoteles, *Categoriae*, c. 5: 4b9-10 (AL 1.3, 92.18-19)

309-317 Cf. Augustinus, *De praedestinatione sanctorum*, c. 15, n. 30 (PL 44, col. 982.11-14) et Anselmus, *De concordia praescientiae et praedestinationis et gratiae Dei cum libero arbitrio*, q. 1, c. 3 (*Opera Omnia* II, ed. Schmitt, Edinburgh 1946, 250.31-251.2; 251.7-14).

321-335 Cf. Aquinas, *Summa theologica* I, q. 14, a. 13 (ed. cit., 83b), et *Scriptum* I, d. 38, q. 1, a. 5 (ed. cit., 912).

329 Aquinas, *Summa theologica* I, q. 14, a. 13 (ed. cit., 83b).

335 Aristoteles, *De interpretatione* I, c. 9: 19a23-24 (AL 2.1, 17.2-3).

351-352 Aquinas, *Scriptum* I, d. 38, q. 1, a. 5 (ed. cit., 914).

Apparatus Criticus

1 d 38 *add in mg* H
1 determinatam] cognitionem et *add* AW
2 quod] arguitur quod AW
2 probatio *om* AW
2 posteriorum] dicitur quod *add* AW
5 scientia determinata *inv* AW
6 contingens[1]] contingenter AW
6 a primo *om* AW
6 contingens[2]] contingenter AW
8 accidit] accidet AW
8 sedebo] sedere AW
8 deus *om* AW
12 sequatur] sequitur AW
12 conclusio] una AW
12 una] altera AW
13 altera] alia AW
14-15 sed... falli *om per homoio* AW
16 scientia determinata *inv* AW
16 limitata est *inv* AW
16 dei] divina AW
17 non] est *add* W
17 determinari] determinate AW
18 determinata *mg* H; limitata H*ac*
19-20 veritas determinata *inv* AW
20 si *om* AW

22 a deo] ab eo AW
23 contingens futurum *inv* AW
24 a nobis *om* W
27 ponitur *mg* H; predicatur H*ac*
29 predicatum] futur *exp* H
33 13] 3 CH
36 primo videndum est *om* W
37 est^{2}] cognitio *add* A
38 ut^{1}] in W
38 ad] in W
39 evenit] et *add* A
42 cognoscere *om* W
44 potest *om* A
46 utrumlibet] utrumque A
52 quam *om* A
57 nec] nunc W
63 omnibus *om* A
67 cognoscibilitate] cognoscente A
73 nec] non W
73 cognoscit] cognovit A
74 et <tum> *om* W; et tunc A
75 cognoscit] cognovit A
77 causet] causat W
77 absolute *om* W
82 tertius articulus opinio thome *add in mg* W
83 oppinio *add in mg* H
83 modo] sic *add* A; W*pc*; sicut W*ac*
84 opinio *om* CH
84 prima *om* CH
84 et *om* ACW
84 scripto] primo *add* AW
85 vel^{1}] uno modo AW
85 vel^{2}] alio modo AW
88-89 hoc... causam^{2} *om per homoio* AW
89 si *om* H
92 remota *om* AW
93 potest contingenter *inv* AW
100 contingenter] contingentes AW
101 ratio... talis *om* W
102 existunt] et *add* AW
108 sit^{1}] cognoscens *exp* H
109 quinque^{1}] tres AW
110-111 vel... successive *om per homoio* C
113 turre] turri AW
114 ista *om* AW
114 uni] uno CH

115 uni nunc *inv* AW
117 aqua vel *om* AW
118 successive... fluvio *om per homoio* C
119 <immensum>] immensus ACHW
119 ideo[1]] ergo AW
119 presens est *inv* AW
120 dicendum *om* AW
121 in actu *om* AW
122 contra hoc *om* AW
122 sub] in AW
123 est... non[2] *om per homoio* A
123 contingentia] rei *add* A
124 actualem[2] *om* CH
125-126 potest contingenter *inv* AW
128 re] rebus AW
128 causata] creata W
129 est *om* AW
132 non-ens] ens AW
132 erit] esset CW
136 dices] dicit A; dicis W
136 est contingens hec] hec contingens A; hoc est contingens W
137 possibile] impossibile CH
142 secundum sensum divisum] pro sensu diviso AW
142 sed] quando *exp* H
143 actualem entitatem *inv* AW; actualem C
144 floritione] arboris *add* AW
145 remota] causa *add* AW
146 scientia dei] floritione arboris CH
146 est simpliciter *inv* AW; est C
146 esse[2]] sed posita floritione *add* C
149 floritione] arboris *add* AW
152 sit *om* CH
152 in se[2] *om* AW
152 ergo] nihil est *exp* H
153 propter] per AW
154 ad essendum simul] simul ad existendum AW
155 quod[1]... cum *om per homoio* W; e *exp* W
156 in illo vel *om* AW
156 presens] scilicet *add* AW
158 suam[1] *om* AW
158 communicare vel *om* AW
159 ens[1] *om* AW
161 quando dicis] quod dicit quod AW
161 quidem] quoddam AW
161-163 eternitatis... partes *om* CH
163 sibi *om* AW

164 quia] quod AW
166 cum] tamen AW
166 nihil] vel H
166 deo] de eo AW
167 haberet] habet AW
169 valent] valunt H
169 tales] talis AW
169 potest] stare in turre et *add* AW
172 doctoris opinio *add in mg* H
172 et *om* AW
173 illorum *om* CH
176 ad *om* CH
176 tamquam... sunt^{2} *om per homoio* A
177 postea] post AW
179 ergo *mg* C
179 ergo uniformiter *inv* AW
180 secundum] secundo AW
182 istud] illud H
182-183 istud... scitum *om per homoio* C
183 notitia] conclusio *add* CH
185 et *om* CH
187 et^{2}] cognito *add* AW
187 quid^{2}] sit *add* AW
189 terminorum] extremorum AW
195 nature vel *om* AW
196 coniunguntur] ut *add* (?) H
196 diameter] diametraliter AW
198 quia] que AW
207 confirmatur quia] propter (?) quod A
208 complexum] connexum A
208 cognoscatur] cognoscitur A
209 ex natura^{1} *om* A
211 necessaria] nomina W
214 futura... existunt *om* W
216 medium] modum A
218 quod] quia A
219 rationem] et *add* A
219 divine *om* W
231 est] esse A
243 intelligens] intellectus A
244 vera] una W
246 modo] medio A
254 formaliter] formalis A
255 albente] albante A
255-256 in... albedinem *iter* A
259 ad primum argumentum *add in mg* H

259 primum argumentum] primam AW
261 didicimus] dicimus AW
263 dupliciter] duplex AW
263 rerum] rei necessarie A; necessario *add* W
263 temporalium] temporalis A
264 est] dicitur AW
265 capitulo *om* AW
265-266 quod... sit] scientiam esse AW
266-267 homo... currere[1]] eam homo scire certitudinaliter A; homo eum scire certitudinaliter W
267 cum] tamen *add* AW
267 currere H*pc*; currente H*ac*
267 sub] cum AW
268 similiter] sic AW
269 ad secundum *add in mg* H
269 dico *om* AW
269 scientia eius *inv* AW
269 dicis] dicit AW
269 a] non exp
270 illud *mg* H
272 ponitur[1]] simul *exp* H
272 ponitur[2]] ponatur AW
273 non *om* AW
273 alterum] alium AW
275 ad tertium *add in mg* H
276 est ista *om* AW
278-279 etiam me *inv* AW
279 non[1] *mg* H
280 fallor] fallitur CH
281 est[2] *om* AH
283 altera[1]] una AW
283 simpliciter *om* AW
283 altera[2]] alia AW
286 et... fine *om* AW
287 ad quartum *add in mg* H
287-288 effectum... unum *om per homoio* A
290 tam] ita AW
290 a[2] *om* AW
291 et... esse[2] *om per homoio* C
292 ut... 39 *om* AW
293 scilicet *om* AW
299 sit *om* AW; nc *exp* A
299 <necessarius>] necessarium ACHW
300 scit[2]] sit CH
301 a *om* AW
302 consiliandum] considerandum W; si<->ndum A

305 si] sed C; *iter* H
306 sciri] scire AW
306 vel a deo] etc AW
306 eveniet] eveniat AW
306-307 eveniet... numquam[1] *om per homoio* C
307 evenit] eveniet AW
309 laboravit] laborant sancti AW
310 predestinationis vel *om* AW
311 ita... peccaturum[2] *om per homoio* C
313 cum] tamen AW
316 oportet] debet AW
316 distinguit] detinguit C; dividit AW
317 et... efficit *om per homoio* A
318 ultimum] ultimam AW
321 thomas] in *add* AW
321 summa *om* AW
321 in parte *om* CH
322 articulo 13 *om* CH
322 et] in AW
326 fore] nisi *exp* H
327 ergo[2]] et *add* AW
330 autem res *om* AW
331 nisi dum] ut W
331 nisi... res *om per homoio* A
331 dum[2]] ut A
331 habet] habent AW
332 erit *om* CH
333 ex] ad A; ab W
337 super *mg* H
338 quia] et *add* H; est *add* A
338 sicut] scitur H
338 et *om* AW
341 necessario[1] vult hoc] vult hoc necessario AW
341 necessario[2] *om* AW
341 ergo[1]] est ergo quod AW
342 est *om* CH
344 est[1] *om* CH
345 illud obiectum *inv* AW
345 sua vel *om* AW
345 vel sui *mg* H
346 illud[1]] quod *add* ACW
346 coexigeret] exigeret AW
349 et *om* H
349-350 et... contingens *om per homoio* C
352 intelligis] intelligit AW
353 ab] in AW

353 et *om* AW
353 intelligis] intelligit AW
354 ita... actu^{2} *om per homoio* C
354 dum^{2} est] de A; de est W
356 suum] actum *exp* H
359 a^{2} *om* AW
362 est] sit AW
366 propositio est *inv* AW
366 ista] ita H; aliter ista *add in mg* H
369-370 ergo... necessarium *om per homoio* C
372 scit] sit H
372 consequens] est *add* AW
372 est possibilis *inv* AW
374 scit^{2}] sit H
375 instans *om* AW; obiectum C
377 esse simul *inv* AW
377 scit^{2} a non] non scit a AW
378 sit] scit AW
379 idem] illud AW

<HENRICUS DE HARCLAY ET AUFREDO GONTERI BRITO *IN I SENTENTIARUM*, DISTINCTIO 39>

Utrum Deus immutabiliter sit prescius futurorum contingentium?

[1] Quod non. Probatio: Sequitur, 'Deus immutabiliter ex parte sui scit A esse; ergo necessario scit A esse'. Consequens falsum. Ergo antecedens. Falsitas consequentis patet ex dictis supra. Probatio consequentie: Omne ens immutabile est formaliter necessarium. Ergo similiter omne verum intellectum, si sit immutabile, est formaliter necessarium. Ergo si Deus scit immutabiliter, non potest non scire.

[2] Preterea, eadem consequentia probatur aliter sic: Omne subiectum quod de necessitate est sub altero contradictoriorum, et non potest mutari de uno in alterum, de necessitate determinat sibi unum illorum -- illa videtur plana, nam non videtur ratio quare {10} non possit mutari nisi quia necessario determinat sibi alterum. Sed Deus de necessitate vel scit A fore vel non scit A fore, quia ista sunt contradictoria, et per positionem non potest mutari de scientia in non-scientiam. Ergo de necessitate de-terminat sibi alterum. Ergo de necessitate scit A futurum vel fore.

[3] Preterea, tertio probatur eadem consequentia sic: "Omne quod est, quando est, necesse est esse," primo *Peryarmenias*. Ergo si aliquid sit in aliquo instanti, non potest non esse pro hoc instanti. Ergo eodem

modo si immutabiliter est pro omni instanti, pro nullo instanti potest non esse. Ergo necesse est esse pro omni instanti.

[4] Preterea, quarto probatur consequentia sic: Omnis potentia ad non-esse alicuius pro aliquo instanti precedit esse illius duratione. Sed si A sit immutabiliter et eternaliter, nulla potentia precedit suum esse duratione. Ergo si sit eternaliter et immutabiliter, nulla est potentia ad eius non-esse pro aliquo instanti. Probatio maioris per auctoritatem multum expressam Anselmi, *De concordia prescientie et predestinationis*, ubi dicit, "Res presens potuit fuisse non presens antequam fuit, et postquam est presens potest postea esse non presens; sed cum est presens, non potest non esse presens." Ergo si sit presens pro hoc instanti, non potest habere in hoc instanti potentiam ut non sit presens in hoc instanti. Si ergo habeat potentiam ut non sit presens in hoc instanti, ista potentia necessario precedit hoc instans. Et sic patet maior.

[5] Preterea, quinto probatur eadem consequentia sic: Si non sequatur, 'Hoc immutabiliter est; ergo hoc necessario stat vel est', oppositum stabit, 'Hoc potest non esse'. Consequens falsum. Probatio: Quidquid repugnat actui, repugnat potentie ad eundem actum. Sed ista repugnant, 'Hoc non est in aliquo instanti', et 'Hoc immutabiliter est'. Ergo potentia ad illum actum repugnabit eidem. Ergo ista repugnabunt, 'Hoc potest non esse', et 'Hoc immutabiliter est'. Ergo oppositum sequitur.

Dicit negando regulam, quia non tenet, 'Cui actui repugnat alia veritas, potentie ad eundem actum repugnabit predicta veritas', nisi illa veritas sit necessitas.

Contra illud, si non repugnant, 'A est in hoc instanti', et 'A potest non esse in hoc instanti', ergo hoc est possibile in hoc instanti, 'A non est'. Ponatur ergo in esse, sequetur impossible. *<Inceptum primae omissionis Aufredi.>*

[6] Sexto probatur consequentia sic: Si Deus immutabiliter <scit> A fore, ergo hec est vera pro omni instanti, 'Deus scit A fore'. Ergo eius contradictoria est falsa pro omni instanti, scilicet, 'Deus non scit A fore'. Quero ergo utrum sit falsa impossibilis aut falsa contingens. Si falsa impossibilis, habeo propositum, quia tunc oppositum est necessarium, scilicet, 'Deus scit A fore'.

Si sit falsa contingens, contra: Illud quod est falsum contingens in hoc instanti, non potest esse verum in hoc instanti. Ergo similiter, quod est falsum contingens pro omni instanti, non potest esse verum pro aliquo instanti. Probatio antecedentis: Magister Obligationum dicit, "Posito falso contingenti de presenti instanti, negandum est idem instans esse." Ergo falsum contingens pro hoc instanti nullo modo potest esse verum in hoc instanti. Ergo eodem modo, si sit falsum contingens pro omni instanti, non potest esse verum pro aliquo instanti.

{50} Magister probat regulam sic: Nam esto quod te esse Rome sit fals-um contingens pro A. Sit autem A hoc instans. Impossibile est quod sit verum pro A, quia si posset verificari, aut ergo per motum aut per mutationem. Non per motum, quia in instanti non est motus. Nec potest verificari per mutationem, quia si esset permutatio ad veritatem in A, tunc esset veritas in A, quia quando est mutatio, tunc est res mutationis. Sic ergo impossibile est ipsum verificari in A. Ergo si ponat ipsam esse veram, oportet eo ipso negare A esse, sicut regula dicit.

Ad principale, quod non est et potest esse, potest incipere esse. Ergo eodem modo, si Deus non scit A esse et potest fore A esse, potest incipere A esse. Sed antecedens est duplex, quia A esse, etsi non sit verum, est possibile. Ergo Deus potest scire A esse. Sed si potest {60} incipere scire A esse, potest mutari. Ergo etc. <*Finis primae omissionis Aufredi.*>

Oppositum: Augustinus, 15 *De trinitate*, capitulo 13, et ponitur in littera, capitulo primo. Et similiter 6 *De trinitate* ultimo: "Cum succed-unt vel redecunt tempora, non succedit vel recedit aliquid de scientia Dei." Ergo est immutabilis.

Ad questionem, dicitur uno modo, et patet superius, quod scientia Dei non solum est immutabilis respectu contingentium, sed necessaria. Nec sequitur propter hoc quod contingens sit necessarium, nam necessitas potest esse in causa remota etsi sit contingentia in causa propinqua, et per consequens etsi sit contingentia in effectu. Exemplum ponitur de sole et floritione arboris, et patet supra. Una auctoritas ad hoc Boetii, quinto *De consolatione*, prosa quinta, dicit, "Fatum ab im-mobilibus providentie proficiscens exordiis, actus fortunasque {70} hom-inum indissolubili causarum connexione constringit." Ergo in Dei provi-dentia omnia sunt determinata et necessaria.

Sed ego teneo oppositum, quod scientia Dei respectu futurorum contingentium non est necessaria, ita quod non solum non scit Anti-christum necessario esse ut necessitas referatur ad adventum Antichristi, immo nec necessario scit Antichristum esse referendo necessitatem ad scientiam Dei.

Hoc probo, primo ostensive, postea ducendo ad impossibile. Primo sic: Impossibile est illud esse simpliciter necessarium quod preexigit contingens in aliquo ordine. Sed Deum scire A fore necessario preexigit Deum velle A fore, quia A fore non est necessarium vel notum ex ter-minis, et ideo dependet ex voluntate Dei. Sed Deum velle A fore est sim-pliciter {80} contingens. Ergo Deum scire A fore non est simpliciter necessarium.

Preterea, effectus qui dependet a causa in fieri et esse et conservari, eo solo habet esse quia causa causat ipsum, et eo solo est necessarius quia causa ipsum necessario causat. Et per consequens, ideo res est contingens quia causa eius totalis, scilicet voluntas divina, contingenter eam causat. Sed sicut voluntas contingenter causat A fore, ita contingenter scientia eius scit A fore.

Secundo, ostendo idem ducendo ad impossibile: Si Deus esset causa remota necessario agens quia volens, sequerentur duo impossibilia vel inconvenientia, primo quod nihil esset contingens, secundo quod nihil omnino esset malum in rebus. Probatio prime consequentie: Illud quod movet quia movetur, si necessario movetur, necessario movet -- illud planum est. {90} Sed Deus est agens et movens necessarium, per te, respectu alicuius effectus propinqui. Ergo ille effectus vel illa res necessario movetur a Deo. Sed in eo quod movetur, movet. Ergo necessario movet. Ergo necessario sequitur effectus. Ergo non contingenter.

Probatio consequentie secunde: Nam agens necessarium agit maximam perfectionem quam potest, et si potest omne malum expellere et bonitatem perfectam inducere, necessario hoc facit, quia est agens necessarium. Sed manifestum est quod Deus volens, potens est omne impedimentum removere, et per te necessario agit. Ergo necessario impedimentum removet, et per consequens omne bonum et nullum malum inducit nec induci permittit.

Preterea, superfluerent omnes cause secunde. Probatio: Nam si causa prima de se esset sufficiens sine causa secunda producere effectum, sicut credimus quod Deus est, si illa {100} causa necessario agit, necessario facit secundum ultimum potentie sue. Et si sic, necessario agit sine causa secunda, quia illa nihil faceret ad productionem effectus.

Dico ergo quod scientia Dei non est necessaria respectu futuri contingentis, et tamen non sequitur propter hoc quod scientia illius sit mutabilis, quia non omne contingens est mutabile, et per consequens nec omne mutabile erit contingens. Probatio istius: Certum est quod pro eodem instanti pro quo A est, pro illo instanti non est mutabilis, et tamen pro illo instanti non necessario est ens, sed contingens. Probatio: Magister in littera, capitulo secundo, dicit quod Deus "potest scire modo, vel prescire, sicut potuit ab eterno scisse vel prescisse." Ergo quando scit, tunc contingenter scit, et tamen pro instanti pro quo scit non potest mutari secundum illud instans.

{110} Preterea, idem Magister, distinctione 43, capitulo ultimo in fine, ubi dicit quod "potest velle Deus quod numquam voluit; nec tamen noviter, sed sempiterna voluntate velle potest; potest enim velle, quia potuit ab eterno voluisse." Ergo plane patet secundum Magistrum quod nec voluntas nec scientia Dei est ex parte sua vel sui necessaria respectu futurorum contingentium, sed tamen invariabilis est.

[~1] Ad argumenta in oppositum, dico ad primum quod non sequitur, 'Hoc est immutabile; ergo necessarium'. Et cum probatur, 'Omne ens immutabile est necessarium', dico ad illud quod non est simile de vero incomplexo et complexo. Et ratio huius est quia verum incomplexum immutabile est tantum verum, et tantum illud est necessarium. Sed non propter immutabilitatem, sed magis econverso, quia est ens necesse esse, ideo est immutabile. Sed {120} verum complexum significat unionem duorum extremorum quorum alterum in se est immutabile et necessarium, alterum immutabile et non necessarium.

Cum dicitur, 'Scientia Dei cadit super A esse, nam scientia Dei in se est invariabilis et necessaria; sed A esse est variabile et mutabile in se, tamen non in comparatione ad scientiam Dei; est tamen contingens in comparatione ad scientiam Dei'. Unde dico quod non omnis potentialitas ad non-esse alicuius est ad hoc quod non esse possit succedere esse, quia in instanti in quo est non potest esse successio ad non-esse. Et tamen in illo instanti non necessario est, quia numquam necessario est, nam voluntas divina pro nullo instanti necessario vult aliquid extra se. Et ideo contingenter vult quidquid preter se vult, tamen non potest mutari.

[~2] Ad secundum argumentum, patet per illud quod maior est falsa, nam alia causa {130} est quare non potest mutari de uno contradictoriorum in alterum quam necessitas essendi sub uno contradictoriorum. Ratio est ista: Quia iste transitus de contradictorio in contradictorium necessario argueret mutationem ex parte scientie Dei, nam res scita non oportet aliter se habere quam prius, nisi quia scientia et actus sciendi aliter se habet quam prius. Sed contingentia respectu unius contradictorii non arguit contingentiam in scientia Dei in se, sed tantum contingentem et non necessariam connexionem inter scientiam Dei et hoc obiectum contingens, et hoc est necessarium.

[~3] Ad tertium, dico quod ista propositio Aristotelis, "Omne quod est, quando est, necesse est esse," est vera ut est de conditionato predicato, scilicet ut hoc totum, 'Est quando est', vel 'est si est', cadat a parte predicati, nam hoc predicatum conditionatum, 'est si est', {140} necessario predicatur de omni ente, sicut hoc predicatum disiunctum, 'fore vel non fore', predicatur de navali bello necessario. Modo hec est fallacia secundum quid et simpliciter, 'Illud quod est, necessario est, si est; ergo hoc nunc necessario est', quia arguitur ab esse conditionato conditione diminuente ad esse simpliciter pro hoc instanti.

Si autem intelligatur propositio conditionalis, tunc esse cadit necessario a parte predicati. Et tunc tota conditionalis est falsa, et est sensus, 'Si hoc est, necessario est', quia antecedens potest esse verum contingens, scilicet, 'Hoc est'. Ideo non valet.

[~4] Ad quartum, nego maiorem. Non enim est necesse quod potentia non-essendi pro aliquo instanti precedat illud instans duratione, nam sicut Deus ab eterno voluit producere mundum contingenter, ita et nunc vult contingenter, et ab eterno voluit.

{150} Ad probationem, dico quod Anselmus ibidem egregie ostendit necessitatem huius propositionis, "Omne quod est, etc," et idem habetur secundo *Cur Deus homo*, capitulo 17, nam dicit ibidem, "Necessitas illa qua necesse est hoc esse presens cum est, est necessitas quod contradictoria non possunt esse simul, et hec est necessitas que sequitur positionem rei in esse, que nihil cogit esse, quia non est causa esse, sed causatur ab esse rei." Et ego concedi quod re posita in esse, et illa stante in esse, non potest non esse et non presens esse. Sed tamen, quia non necessario est presens in esse, ideo potest removeri ab esse, non per mutationem, sed tantum per potentiam precedentem actum natura, et non duratione. Et ista non posita vel remota ab esse, alterum potest poni in esse.

[~5] Ad quintum dico quod regula non valet sicut fuit dictum.

{160} Ad probationem, cum dicis, 'Non repugnat potentia, ergo compossibilis est; ponatur ergo in esse, non accidet impossibile', dico quod propter eius positionem non accidit impossibile, sed propter hoc quod simul ponitur in esse cum alia que sibi contradicit, unde resultat nova incompossibilitas. Verbi gratia, possibile est hoc corpus esse in A instanti in hoc loco. Possibile est similiter aliud corpus esse in hoc eodem loco in eodem instanti. Ponatur ergo in esse, 'Hoc est in hoc loco', et nullum impossibile. Eodem modo, si illud corpus poneretur in hoc loco, nullum accideret impossibile. Tamen, stante hoc corpore in hoc loco et manente, alterum non potest poni cum eo in hoc loco. Unde non sequitur, 'Non habent potentiam essendi simul in loco uno; ergo non habent simul potentiam essendi in uno loco', nam simultas potentie non arguit potentiam simultatis. *<Inceptum omissionis secundae Aufredi.>*

{170} [~6] Ad sextum, dico quod hec est falsa pro omni instanti. Et cum arguis, 'Aut falsa contingens aut falsa impossibilis', dico quod falsa contingens.

Ad probationem, dico quod regula Magistri non tenet, "Posito falso contingenti etc."

Ad probationem, dico quod veritas potest esse in hoc instanti, et tamen nec per motum vel per mutationem. Et ratio huius: Nam veritatem posse esse in hoc instanti, puta, 'Tu es Rome', non arguit nisi non-repugnantiam huius instantis ad te esse Rome. Et non plus hoc instans repugnat quin in eo sit hec veritas, 'Tu es Rome modo' -- sicut non repugnavit antequam esset instans (verbi gratia, si hoc corpus est in loco nunc, non repugnat huic loco pro nunc habere aliud corpus) -- quam repugnavit prius ante hoc instans, quia repugnantia est ratione forme que una et eadem est sive sit hic sive non. Sed tamen repugnat huic corpus esse simul tempore cum {180} alio in eodem loco. Et ista repugnantia non est ratione instantis, sed ratione duorum incompossibilium pro eodem instanti. Unde ad hoc quod aliquod instans contingenter respiciat aliquando verum, vel sit mensura alicuius veri, sufficit quod sit prior illo natura, sicut superficies colore, et quod non sit necessario connexio ex natura rei inter illa, sicut nec inter superficiem et album. Sed ad hoc quod mutatio fieret, quia extrema mutationis sunt privatio et habitus, ideo non potest esse in eodem instanti.

Preterea, ipsemet Magister obicit contra se vel contra regulam tripliciter. Quarum ultima ratio, que est ad propositum, est illa: Probat Magister quod, si regula sit bona, quod sequitur inconveniens. Probatio: Hec est falsa de presenti instanti, 'Tu es Rome'. Ponatur ergo ista, 'Tu es Rome'. Deinde propono, 'A fuit venturum'. Sit A presens instans. Hec est {190} concedenda, quia vera et impertinens. Deinde propono, 'A fuit'. Hec est neganda, quia falsa et non sequens ex posito nec ex concesso. Ergo eius opposita est concedenda, 'A non fuit', et hoc est verum, quia modo est, et prius non fuit. Deinde propono, 'A non est'. Hec est concedenda, quia sequens ex posito. Sequitur enim secundum regulam, 'Tu es Rome; ergo A non est', quia "posito falso contingenti etc." Deinde propono, 'A erit'. Aut concedis aut negas. Si concedas, concessisti impossibile possibili facta positione, quia hec est impossibilis, 'Presens instans erit', quia quod est et erit, necessario est in diversis partibus temporis. Hoc est impossible de instanti. Si tu negas, negas consequens et concedis antecedens; ergo male. Probatio istius: Sequitur enim, 'Si tu es Rome et A fuit venturum et non fuit, nec est, ergo erit'.

Respondet Magister dicens quod proposita ista, 'A fuit', hec est concedenda, quia {200} sequens ex posito, nam sequitur, 'Si tu es Rome et A non est et fuit venturum, et non erit, ergo fuit'. Sed tamen videtur mihi quod non sufficit ista ratio. Probatio, arguo sic: Quando due forme incompossibiles in eodem subiecto, tamen nate esse in illo vel possibilies esse in illo, puta albedo et nigredo circa parietem, etsi existentia unius in subiecto arguit absentiam alterius, tamen non arguit absentiam subiecti communis. Sed hoc instans ex parte sui est communis ad te esse Rome et te esse Novarie, nam non repugnat huic instanti quod tu sis Rome in illo, etsi repugnet existentie tue Parisius quod tu sis Rome vel Novarie in eodem instanti, sicut superficiei pro quocumque instanti non repugnat albedo, etsi nigredini repugnaret albedo in eadem superficie. Ergo ex hoc quod ponitur te esse Rome nunc, non propter hoc debet negari hoc nunc esse, sed debet negari oppositum et incompossibile huius esse in hoc nunc.

Propterea, dico quod regula non est contra me. Probatio: Aliud est ponere propositionem esse veram que est falsa de presenti instanti, et aliud ponere falsam esse veram pro hoc instanti. Nam ponendo propositionem que est nunc falsa esse veram nunc, non accideret impossibile. Sed ponere falsam nunc esse veram nunc est ponere impossibile.

Ad propositum, dico quod posito falso contingenti quod est falsum pro hoc instanti, non propter hoc negandum est illud instans esse, nam sic ponendo, ponere illud quod est nunc falsum esse verum nunc et non accideret impossibile, quia falsum, quod nunc est falsum, non necessario est nunc falsum. Ideo nunc potest esse verum sine corruptione huius instantis. Sed ponere propositionem falsam de presenti instanti, stante falsitate eius pro presenti instanti, si ponerem eam esse veram, necessario oporteret eam ponere pro <alio> instanti, quia stante falsitate pro hoc instanti, non potest simul cum hoc poni veritas in eodem instanti. Ideo stante illa, si ponatur veritas, oportet ponere pro alio instanti <et> ergo ponere aliud instans esse, et per consequens hoc instans non esse. Et ego concedo quod dum est falsa pro hoc instanti, non potest simul cum hoc esse vera in eodem. Si ergo ponatur vera, oportet referre ad aliud instans.

Apparatus Fontium

14-15 Aristoteles, *De interpretatione* I, c. 9: 19a23-24 (AL 2.1, 17.3-4).

22-24 Anselmus *De concordia... gratiae*, q. 1, c. 2 (ed. cit., 249.19-23, non verbatim).

46-56 Cf. Guilelmus de Shirwode, *De obligationibus* (ed. Romuald Green, "The Logical Treatise *'De Obligationibus'*: An Introduction with Critical Texts of William of Sherwood and Walter Burley," unpublished, par. 1.26).

61-62 Cf. Augustinus, *De trinitate* XV, c. 13, n. 22 - c. 14, n. 23 (CCL 50a, 495-496). Cf. Lombardus, *I Sent.*, d. 39, c. 1 (ed. cit., 280.16-281.11).

62-63 Augustinus, *De trinitate* VI, c. 10, n. 11 (CCL 50, 241.25-26).

64-68 Cf. Aquinas, *I Sent.*, d. 38, q. 1, a. 5 (ed. cit., 909-910).

68-70 Boethius, *De consolatione Philosophiae* IV, prosa 6 (CCL 94, 80-81.76-79).

106-107 Lombardus, *I Sent.*, d. 39, c. 2 (ed. cit., 281.33-34).

110-112 Lombardus, *I Sent.*, d. 43, c. unico (ed. cit., 303.15-17).

150-154 Anselmus, *De concordia... gratiae*, q. 1, c. 2 (ed. cit., 249, non verbatim). Cf. Anselmus, *Cur Deus homo* II, c. 17 (*Opera Omnia* II, ed. Schmitt, 125).

188-198 Cf. Shirwode, *De obligationibus* (ed. cit., par. 1.29-30).

199-201 Cf. Shirwode, *De obligationibus* (ed. cit., par. 1.33).

Apparatus Criticus

1 distinctio 39 *add in mg* H
1 distinctio 39 *add* H
1 contingentium] probo *add* AW
2 probatio *om* AW
4 formaliter] simpliciter AW
5 intellectum] in intellectu AW
11 scit^{1}] sit H
11 a^{1}] futurum *exp* H
11 non scit a] scit a non AW
11 scit^{2} *s.l.* H
12-13 determinat... necessitate *om per homoio* C
13 futurum vel *om* AW
15 aliquo] hoc AW
15-16 non^{1}... instanti^{1} *om per homoio* CW
16-17 hoc... ergo *om per homoio* A
19-20 immutabiliter... sit *om per homoio* CH
19-20 nulla... eternaliter *om per homoio* A
21-22 multum expressam *inv* W, A*ac*
24 non esse *inv* AW
24 pro] in AW
24 pro hoc instanti *mg infra* H
28 preterea *om* AW
29 hoc^{1} *om* AW
29 stat vel *om* AW
34 tenet] si *add in mg* A
35 predicta] eadem AW
36 si] ista *add s.l.* A
37 est^{1} *om* W; *s.l.* A
37 est^{2}] esse AW
37 sequetur] sequitur AW
39 <scit>] sit CH
42 habeo] habes C
51 sit autem *iter* H
51 posset] esse nullum *exp* H
57 ad principale *add in mg* H
61 contra *add in mg* H
62 succedit] succedunt C; *corr in mg ex* succedunt H
63 recedit] recedunt CH
64 ad questionem *add in mg* H
64 et] est vel *add* CH
66-67 in^{2}... contingentia *om per homoio* C
68 et... quinto] pro isto boetius libro AW
69 proficiscens] prosiciscens H; proficisionis C

70 providentia *corr s.l. ex* prodentia H
71 determinata et *om* AW
72 quod scientia dei respectu futurorum contingentium non est necessaria secundum scotum *add in mg* H
74 adventum] eventum AW
74 immo *iter* C
74 scit] sit CH
74 esse *om* CH
76 postea] secundo AW
77 esse simpliciter *inv* AW
78 necessarium vel *om* AW
78 vel notum *mg* H
80 non *om* CH
80 simpliciter *om* CH
81 causa *iter* H
85 scientia eius *inv* AW
86 esset *om* A; ut *s.l.* A
86 remota] est *add s.l.* A
87 sequerentur] sequentur AW
87 impossibile vel *om* AW
89 illud planum est] hoc est planum AW
90 agens et movens *inv* AW
91-92 movet... necessario *om per homoio* A
93 consequentie secunde *inv* AW
98 omnes cause *inv* AW
98 causa prima *inv* AW
98 de se *om* AW
99-100 quod... facit] si ista causa necessario causat W; si causa necessario causat A
101 agit] causat AW
103 illius] dei AW
104 et... mutabile^{2} *om per homoio* C
104 nec] non AW
104 erit] est AW
106 magister *add in mg* H
107 scire modo *inv* AW
107 vel prescire *om* AW
107 scisse vel prescisse] scire vel prescire AW
110 ubi *om* AW
111 potest velle deus] deus potest velle AW
111 nec *mg* H; et *exp* H
113 est *om* AW
113 sua vel *om* AW
113 necessaria] est *add* AW
115 ad argumentum in oppositum ad primum *add in mg* H
115 ad^{1}] sed ad AW

115 oppositum] respondendum *add* A; respondendum est *add* W
115 dico ad primum] ad primum dico AW
119 ens] esse *exp* AW; non *add s.l.* A
120 significat] signat AW
120 duorum *om* AW
120 alterum *om* CH
121 alterum] est *add* AW
121 immutabile] mutabile CH
122 dei^{2} *om* CH
122 in se est] est in se AW
124 est... dei^{2} *om per homoio* ACW
126 successio] succedere AW
126 in^{2} *om* AW
129 ad secundum *add in mg* H
129 per illud *om* AW
130 alterum] alio CH
131 iste *om* AW
133 nisi... prius^{2} *om per homoio* CH
135 tantum] necessarium *exp* AW
137 ad tertium *add in mg* H
139 si est^{3}] sicut C; sive AW
141 predicatur *om* AW
144-145 conditionalis... tota *om per homoio* A
146 contingens] consequens AW; falsum *add in mg* A
147 ad quartum *add in mg* H
149 et^{1} *om* AW
149 nunc] tunc AW
150 ad probationem anselmi *add in mg* H
151 secundo] libro *add* AW
152 quod] qua AW
154 concedi] concedo ACW
154-155 rei... esse^{2} *om per homoio* C
155 et^{1}... esse^{2} *om per homoio* AW
155-156 non^{4}... presens] presens non est necessario AW
157 vel *s.l.* H
159 ad quintum *add in mg* H
159 fuit dictum *inv* AW
160 cum dicis] convenit A; cum dicitur W
161 accidit] sequitur AW
161-162 dico... impossibile *om per homoio* C
163 incompossibilitas *corr in mg ex* impossibilitas H; impossibilitas C
164 aliud] illud AW; istud aliter aliud C; istud A; aliter aliud *add in mg* H
165-166 et... loco^{1} *om per homoio* AW
166 accideret] accidet AW
167 loco *om* AW

167-168 habent... non *om* A; habent simul potentiam in uno loco ergo habent esse *add in mg* A
168 loco^{1}... in^{2} *om per homoio* W
168 habent... essendi *om* A; habent potentiam essendi simul *exp* A; simul *add* A
170 ad sextum *add in mg* H
170 instanti] hec non est *add* CH
171 impossibilis *corr s.l. ex* possibilis H
178 repugnavit] repugnaret C
178 quia] sed C
181 instanti] va sive sint huius sive non sed tamen repugnat cat *exp* H
187 quod^{1}] quia C
188-189 ponatur... rome *om per homoio* C
194 concedas] concedis C
200 posito] proposito H
205 non *mg* H; si *exp* H
207-208 etsi... albedo *om per homoio* C
208 non] ponitur *exp* H
209 nunc^{1}] non C
210 propterea] preterea C
212 nunc^{2} *mg* H
213 nunc^{2}] non C
215 ponendo] posito C
217 nunc^{2}] non C
219 <alio>] aliquo CH
220 hoc^{1} *mg* H; aliquo *exp* H
223 ponatur] simul *add* CH
223 oportet] oporteret C

De Memoria Artificiali:
Time and Memory in the Thought of Thomas Bradwardine

Edith Wilks Dolnikowski

With his wide-ranging interests in mathematics, natural philosophy and theology, Thomas Bradwardine was one of the most innovative thinkers of the late middle ages. His remarkable ability to integrate into his theological arguments the insights he derived from his study of natural philosophy gave his writings uncommon force and conviction. Nowhere is his talent for synthesis more evident than in his treatment of time, which combines Aristotelian definitions of motion, continuity and sequence with Augustinian conceptions of infinity and eternity. Given his strong advocacy of Augustinian theology, it is somewhat surprising that Bradwardine did not emphasize in his own treatments of time Augustine's theory that memory is the means through which human beings perceive the passage of time. Although Bradwardine's view of time, as he expressed it in such texts as *De proportionibus*, *De continuo*, *De futuris contingentibus* and *De causa Dei* does not explicitly rely on Augustine's treatment of memory, it is consistent with Augustine's psychological approach to time.

Perhaps Bradwardine chose not to include explicit material about memory in his studies of time because he had already explored the psychological dimensions of memory separately in the short text *De memoria artificiali*. Some questions have been raised about the attribution of this text to Bradwardine. James Weisheipl, working from a corrupted manuscript, was uncertain about whether Bradwardine composed this text.[1] Subsequently, better versions have been discovered which strengthen the attribution.[2] In *The Book of Memory: A Study of Memory in Medieval Culture*, Mary Carruthers makes a strong argument in favor of Bradwardine's authorship based on internal evidence: first, "[t]he scholastic form of the treatise indicates a university setting for both the author and his audience;" second, mathematical references within the text "suggest someone, like Bradwardine, who was familiar with university-style arithmetic;" and third, the mnemonic devices outlined in the text make use of examples which refer to English political life in the mid-1330s.[3] I would argue that the attribution is even stronger when one evaluates this text from the perspective of Bradwardine's treatment of time in his other writings. Indeed, I would contend that the treatment of memory in *De memoria artificiali* is consistent with the view of time that Bradwardine advanced in other texts and serves to underscore his reliance on Aristotle and Augustine as authorities, in matters both of theology and human psychology.

In this essay I shall examine the approach to memory presented in *De memoria artificiali* by highlighting the Aristotelian and Augustinian elements within the text that correspond to Bradwardine's use of these concepts in his treatment of time. I will also examine briefly the stylistic elements of *De memoria artificiali* that are also present in Bradwardine's other writings. In addition to verifying the authorship of the treatise, these comparisons will help us to explain Bradwardine's tendency to treat memory as a subject related to but not directly dependent on a psychological understanding of time.

Although Bradwardine did not compose, so far as we know, a separate treatise about time, his view of time is discernible from his treatises about various mathematical, philosophical and theological topics. In his works of natural philosophy, he advanced the Aristotelian definition of time as the number or measure of motion. He assumed that time is an infinitely divisible continuum moving in a single direction from past to future. Within this larger continuum, smaller continua, such as the motion of an object, can be observed and measured. This understanding of time informed his treatments of proportionality, continuity and acceleration.[4]

Like Aristotle, Bradwardine was also intrigued by the philosophical implications of this definition of time. In *De futuris contingentibus*, for example, Bradwardine posed the question: how can future events be truly contingent when they exist already completed in the mind of God? Here Bradwardine concluded tentatively that the human experience of time as a sequential contingent continuum must be fundamentally different from God's perception of time.[5]

Bradwardine's great theological treatise, *De causa Dei contra pelagium*, expresses his mature view of time. Drawing on Augustinian and Boethian discussions of time and eternity, Bradwardine developed a sophisticated solution to the problem of time. From a human perspective, he argued, time is a sequential continuum with Aristotelian characteristics. From a divine perspective, by contrast, time is merely a reflection of eternity, in which God sees all of creation in a single moment. Because God's perception of creation is fundamentally different from human perceptions, human beings are not constrained in their choices by divine foreknowledge. The future remains contingent for us because God has placed us in a world governed by the sequential, infinitely divisible continuum of time.[6]

Bradwardine's treatment of memory reflects a similar tendency towards integration of physical, psychological and theological principles. The *De memoria artificiali*, which Bradwardine presumably addressed to a university audience, assumes that the reader wants to learn a formal method for enhancing memory. He begins the treatise, therefore, by describing the nature of memory in terms of Aristotelian natural philosophy:

> For trained memory, two things are necessary, that is, firm locations and also images for the material; for the locations are like tablets on which we write, the images like the letters written on them; moreover, the locations are fundamental to the images. ... With regard to these locations, then, six matters are distin-guished, that is, size, configuration, characteristics, number, order and distance-away.[7]

Bradwardine typically draws on the language of geometry to evaluate these features of memory.[8] He asserts, for example, that "a location's config-uration should be like a four-sided oblong."[9] Each location within the memory must then be organized to maximize the retrieval of information:

> [O]rder requires that the places have contiguity and direction, so that memory may with facility find all the inscribed images in their places easily, in forward order or backwards. Concerning the depiction of the background within the locations [locorum distanciam] two things are determining, that is extent and detail [qualitas]. ... With regard to the rememberer who mentally perceives the locations, three things are required for effectiveness, that is, a clear and firm habitual memory of these locations. ... Secondly, that one should know promptly without counting whichever number in the order of places it may occupy. Thirdly, when one must use one of the locations for a memory task, one imagines oneself positioned at the optimal distance from it and from there reviews the location.[10]

Bradwardine describes the images in similar terms, stating that each image has "size, quality, order, and number."[11]

In this way, Bradwardine constructs a theory of memory according to the principles of Aristotelian natural philosophy. Individual memories, or images, become objects which can be organized and stored in the mind. Such concepts as continuity, contiguity, order and sequence, which define natural phenomena, can also be used to define mental objects. Although in an Aristotelian sense, memory is not the same as time, memory and time share some of the same qualities. Moreover, because memories are sequences of images that are stored and retrieved in time, time can be seen as the larger continuum in which these smaller sequences exist and are measured. To the extent that it incorporates concepts such as proportion, continuity, number and quality, Bradwardine's construction of memory reflects a strong dependence on the principles of Aristotelian natural philosophy.

In one respect these Aristotelian principles proved inadequate to explain the nature of memory. Aristotle acknowledged that human beings make use of sensory observation to perceive the passage of time but he applied his analysis of perception more to the problem of time than to the problem of memory. Aristotle argues in Book IV of his *Physics*, for example, that:

> ...we perceive movement and time together; for even when it is dark and we are not being affected through the body, if any movement takes place in the mind we at once suppose that some time has indeed elapsed; and not only that but also, when some time is thought to have passed, some movement also along with it seems to have taken place. Hence time is either movement or something that belongs to movement.[12]

Sensory observations, then, help us to perceive that we exist within the continuum of time and help us to measure how much time has passed. Aristotle's association of time and motion, however, does not take into account the way that human beings perceive the continuum of past, present and future in their conscious minds.

In addition to the Aristotelian assumption that memory has size, number, quality and sequence, therefore, the *De memoria artificiali* also incorporates Augustinian assumptions about human psychology. According to Augustine, time and memory are interrelated because memory is a component of the cognitive process human beings use to perceive the passage of time. He argues in the *Confessions*, for example, that "there are three times, a present of things past, a present of things present, a present of things future. For these three exist in the mind, and I find them nowhere else: the present of things past is memory, the present of things present is immediate intuition, the present of things future is expectation."[13] Augustine goes on to illustrate these components of time with the example of reciting a psalm:

> Suppose that I am about to recite a psalm that I know. Before I begin, my expectation is directed to the whole of it; but when I have begun, so much of it as I pluck off and drop away into the past becomes the matter for my memory; and the whole life of the action is stretched out between my memory, in regard to what I have said, and my expectation, in regard to what I am still to say.[14]

Bradwardine's theory of memory is based on this Augustinian notion that memory is a component of time. To retrieve a memory successfully, Bradwardine argues, one must be attentive to the sensory images that bring the past back to the present:

> Each location [within the memory] should be of moderate size, as much as one's visual power can comprehend in a single look, such as a small garden or arcaded space [like a cloister]. Indeed, memory is most powerfully affected by sensory impression, most strongly by vision; wherefore something occurs in memory as it customarily occurs in seeing.[15]

Thus Bradwardine assumes, as Augustine did, that sensory perception is a vital component of our cognitive ability to distinguish between past, present and future.

The trained memory takes advantage of this ability to perceive to organize material in the mind so that it can more easily be recalled. Bradwardine recommends associating strong perceptions with material to be remembered because strong perceptions lead to stronger memories:

> [the quality of the material] should be wondrous and intense, because such things are impressed in memory more deeply and are better retained. However such things are for the most part not average but extremes, as the most beautiful or ugly, joyous or sad, worthy of respect or something ridiculous for mocking, a thing of great dignity or vileness, or wounded with great open wounds with a remarkably lively flow of blood, or in another way extremely ugly, strange of clothing and all bizarre of equipment, the color also very brilliant and intense, such as intense, fiery red, and the whole color strongly altering the appearance. The whole image also should have some detail or movement, that thus more effectively than through what is routine or at rest, they may be commended to memory.[16]

A trained memory cannot depend on the intensity of the material alone, however. One must also consider carefully how to arrange these intense images in the mind:

> With respect to the images, moreover, two things are helpful for the rememberer. First, he should not make his images for himself too quickly and inquisitively, but a little at a time he imprints whatever portion he wants for himself profoundly and stablely by cogitating. The second thing is that not just the image itself but equally its order with relation to what most immediately precedes and what follows he should equally diligently commit to his memory, so that wherever he may wish, forwards or backwards, he may recollect easily regarding this one particular.[17]

Again, Bradwardine assumes the same continuum of memory, present intuition and expectation found in Augustine's treatment of time. In fact, Bradwardine argues, the continuum of time is not simply a neutral force of nature because the continuum of time can be actively and effectively used to expand our perceptions of nature. By consciously organizing our memories in the present, we make them available for future use. Our minds allow us to move back and forth through the continuum of time in a way that our physical bodies cannot.

Although *De memoria artificiali* does not specifically mention the concept of time, it presents a theory of memory which is entirely consistent with Bradwardine's view of time. From the tradition of Aristotelian natural philosophy we find references to the size, quality, number and sequence of the components of memory. From the Augustinian tradition we find references to sensory perception as critical factor in distinguishing between past and present. As is typical of many of Bradwardine's works of natural philosophy and theology, he incorporated elements of Aristotelian and Augustinian thought that supported his view. In the specific case of memory, Bradwardine assumes that memory is a continuum within the mind that reflects the natural continuum of time. Aristotle is correct that time is the number or measure of motion which we can observe through our senses. Because we can observe the passage of time and store our observations in our memory for future use, however, we perceive time according to Augustine's continuum of past memory, present intuition and future expectation.

The presence of these arguments and assumptions, which are selected and woven together with considerable care, serve to confirm Bradwardine's authorship of *De memoria artificiali*. Its highly organized structure and use of mathematical terms further strengthens this attribution. Students of Bradwardine's natural philosophy and theology are fortunate to have access to this brief text; for although it does not add anything strikingly new about his thought in general, it provides fascinating insights into his understanding of human cognition, perception and memory.

The Church of Our Saviour, Brookline, Massachusetts

Notes

1. For Weisheipl's evaluation of *De memoria artificiali* and other works by Bradwardine, see "Repertorium Mertonense," *Mediaeval Studies* 31 (1969), 174-224.

2. For additional information on the text see Beryl Rowland, "Bishop Bradwardine, the Artificial Memory, and the House of Fame," in *Chaucer at Albany*, ed. Russell Hope Robbins (New York: Burt Franklin, 1975), 41-62; and Heiner Gillmeister, "An Intriguing Fourteenth-Century Document: Thomas Bradwardine's *De arte memorativa*," *Archiv* 220 (1983), 111-4.

3. (New York: Cambridge UP, 1990, 350). I use Carruthers' translation of *De memoria artificiali*, 281-8.

4 See Edith Wilks Dolnikowski, *Thomas Bradwardine: A View of Time and a Vision of Eternity in Fourteenth-Century Thought* (New York: Brill, 1995), 73-131.

5. *Ibid.*, 131-64.

6. *Ibid.*, 165-207.

7. Carruthers, 281.

8. For a review of Bradwardine's application of geometric analysis to philosophical and theological problems, see Dolnikowski, 123-31, 180.

9. Carruthers, 281.

10. *Ibid.*, 282.

11. *Ibid.*

12. *Physica*, iv.11 219a 4-14; trans. Jonathan Barnes, *The Complete Works of Aristotle: The Revised Oxford Translation*, vol. I (Princeton: Princeton UP, 1984), 371.

13. *Confessions*, XI.XX.26, trans. James McEvoy, "St. Augustine's Account of Time and Wittgenstein's Criticisms," *Review of Metaphysics* 37 (1984), 557.

14. *Confessions*, XI.XXVIII.38, trans. McEvoy, 562.

15. Carruthers, 281.

16. *Ibid.*, 282.

17. *Ibid.*, 283.

"Thirty Days Hath September": Oral Rendition of Time in Sixteenth-Century England

J.D. Alsop

Historians of early modern England have devoted attention to popular conceptions of time in the terms of the celebratory year or seasonality of activity.[1] There exists, as well, investigation of contemporary attitudes to the life cycle, albeit generally drawn from, and related to, the literate minority.[2] The issues of popular knowledge of the calendar appears to have generated less interest, at least for sixteenth-century England. Although the following piece of evidence was undoubtedly committed to paper by an anonymous literate official, it can stimulate thought on how well-known versions of the "Thirty days hath September" verse were within the popular culture of this society.

An officer of the Elizabethan Exchequer of Audit at Westminster compiled a reference notebook for sundry official business: the dates and rates for parliamentary taxation; the customs rates on imports and exports of wine and wool; the values of large amounts of small coins in pounds, shillings, and pence (i.e., 4,000 pence = £ 16.13.4); the French words for figures from one to one thousand; and so on.[3] The notebook was begun about the year 1566, with additions up to 1594. One page contains a version of the days in the month verse:

> Thirty dayes hath September
> April June and nouember
> ffebruary hat 28 a lone,
> and all the rest thirty and one.

Would the act of committing the verse to paper indicate the rarety of the verse in Tudor society?

McMaster University

Notes

1) David Cressy, "The Protestant Calendar and the Vocabulary of Celebration in Early Modern England," *Journal of British Studies* 29 (1990), 31-52; C. Phythian Adams, "Ceremony and the Citizen: The Communal Year at Coventry 1450-1550," in P. Slack and P. Clark, eds., *Crisis and Order in English Towns* (London, 1972).

2) J.A. Burrow, *The Ages of Man* (Oxford, 1986); Samuel Chen, *The Pilgrimage of Life* (London, 1962); Keith Thomas, "Age and Authority in Early Modern England," *Proceedings of the British Academy* 62 (1976), 205-48.

3) Public Record Office, E 179/272/4 (unfoliated).

Editorial Policy

Disputatio is an annual journal of the late middle ages which publishes articles *only* on the subject of the announced theme for each volume. Preferred length for scholarly articles is 4,000-12,000 words and for research notes is under 2,500 words. Review essays will be assigned by the editors; please contact us if you might wish to be assigned a topic. We are also very committed to publishing bibliographies; please contact us if you are interested in compiling one. For review essays, inquiries from younger scholars or those working on dissertations in the field are particularly welcome. Although the preferred language of *Disputatio* is English, we will consider articles in all major European languages and Latin. All articles not in English must be accompanied by a substantial abstract in English. Manuscripts must conform to MLA style. Although contributors are requested to use hard copy for initial submission of manuscripts, final copies of accepted manuscripts must be submitted on disk (Windows *only* -- we *cannot* use Macintosh disks). Authors should send three copies of the manuscript and a stamped self-addressed envelope when submitting work. All articles will receive blind review by specialists in the field; final decisions concerning publication, however, rest exclusively with the editors.

Topics for Subsequent Volumes

1998: Transformation, Translation, & Transubstantiation. Deadline: 1 Nov. 1997.
1999: Grammars and Grammarians in the Late Middle Ages. Deadline: 1 Nov. 1998.
2000: Late Medieval Disputations. Deadline: 1 Nov. 1999.

Address all editorial correspondence to:

North American contributions:
Carol Poster, Editor, *Disputatio,*
English Department, Wilson 2-176,
Montana State University,
Bozeman MT 59717-0230
e-mail: poster@english.montana.edu

All other contributions:
Richard J. Utz, Editor, *Disputatio*,
Englisches Seminar,
Universitadt Tuebingen, Wilhelmstr. 50,
D-72076 Tuebingen, Germany
e-mail: Richard.Utz@Uni-Tuebingen.De

Address subscription inquiries to:

Northwestern University Press
Chicago Distribution Center
11030 South Langley Avenue
Chicago IL 60628
Phone: 800-621-2736 or 312-568-1550
Fax: 800-621-8476 or 312-660-2235